THE WASHINGTON NAVY YARD

THE WASHINGTON NAVY YARD
An Illustrated History

Edward J. Marolda

with Foreword by
Rear Admiral Christopher E. Weaver, USN
Commandant Naval District Washington

NAVAL HISTORICAL CENTER

Washington, 1999

**Secretary of the Navy's
Advisory Subcommittee on Naval History**

Dr. David A. Rosenberg, *Chairman*

CDR Wesley A. Brown, CEC, USN (Ret.)
Dr. Frank G. Burke
J. Revell Carr
VADM Robert F. Dunn, USN (Ret.)
VADM George W. Emery, USN (Ret.)
Dr. Jose-Marie Griffiths
Dr. Beverly Schreiber Jacoby
David E. Kendall
Honorable G. V. Montgomery
Dr. James R. Reckner
Dr. William N. Still Jr.
ADM William O. Studeman, USN (Ret.)
Virginia S. Wood

Library of Congress Cataloging-in-Publication Data

Marolda, Edward J.
    The Washington Navy Yard : an illustrated history / Edward J.
    Marolda ; with foreword by Christopher E. Weaver
        p.   cm.
    Includes bibliographical references and index.
    ISBN 0–945274–41–6 (pbk. : alk. paper)
    1. Washington Navy Yard—History. I. Title.
VA70.W3M37 1999
359.7'09753—dc21   99–31616

♾ The paper used in this publication meets the requirements for permanence established by the American National Standard for Information Sciences "Permanence of Paper for Printed Library Materials" (ANSI Z39.48–1984).

For sale by the
United States Government Printing Office,
Superintendent of Documents, Mail Stop: SSOP,
Washington, DC 20402–9328
ISBN 0-16-050104-0

Contents

List of Illustrations *vii*

Foreword *ix*

Preface *xi*

Acknowledgments *xiii*

The Early Years *1*

The War of 1812 *9*

Supporting the New Navy *11*

The Father of Naval Ordnance *15*

Civil War Naval Arsenal and Operating Base *21*

The Post-Civil War Era *35*

Arming a New Navy *39*

Ordnance for a Navy Second to None *53*

Between the World Wars *59*

Ordnance Nerve Center for a Global Conflict *69*

A New Mission for the Navy Yard *75*

Notes *101*

Bibliography *103*

Index *107*

List of Illustrations

Cover: Lithograph of the Washington Navy Yard, published during the Civil War

The Early Years
"Captain Thomas Tingey" *xiv*
Triopoli Monument *1*
"Benjamin Stoddert" *2*
Tingey House, or Quarters A *3*
"USS Chesapeake" *4*
"Mosquito Squadron" *5*
Quarters B, the oldest structure *6*
Building 1, the old Commandant's Office *7*
"USS Constellation" *8*

The War of 1812
"Joshua Barney" *9*
"United States Ship-of-the-Line Columbus" *10*

Supporting the New Navy
"United States versus HMS Macedonian" *11*
"President John Quincy Adams" *12*
"USS Brandywine" *12*
"The Marquis de Lafayette" *13*
"USS Water Witch" *14*

The Father of Naval Ordnance
"Awful Explosion of the 'Peacemaker'" *15*
Interior view of the experimental battery *16*
Civil War 9-inch Dahlgren gun *16*
Commodore Charles Morris *17*
John A. Dahlgren *18*
Washington Navy Yard waterfront, 1866 *19*
The first Japanese delegation to visit the United States *20*
U.S. Marine detachment of the navy yard garrison *21*

Civil War Naval Arsenal and Operating Base
"Honorable Gideon Welles, Secretary of the Navy" *22*
The 71st New York Infantry Regiment *22*
"The Contest for Henry Hill" *23*
Civil War lithograph shows the navy yard's two shiphouses *25*
President Abraham Lincoln *26*
African-American sailors mend their clothes *27*
Colonel Ulric Dahlgren *28*
"The Ironclads" *28*
Hand-tinted woodcut of fishermen along the Eastern Branch *29*
The gun park of the navy yard *30*

The main ordnance foundry *30*
Brazilian steamer *Paraense* *31*
Ordnance stores and armor test pieces *31*
John Wilkes Booth *32*
Monitor *Montauk*, autopsy site *33*
Confederate naval officer Raphael Semmes *34*

Post-Civil War Era
Lieutenant William N. Jeffers *35*
John Surratt's arrival at the navy yard *36*
Gunboat *Nipsic* *37*
The new steel ship USS *Atlanta* *38*

Arming a New Navy
Washington Navy Yard, 1888 print *39*
Naval Gun Factory workers *40*
Sailors work on gun components *40*
Workers of the pattern and joiner shop *41*
Construction of the Experimental Model Basin *41*
Rear Admiral David W. Taylor *42*
Filling the Experimental Model Basin *42*
Civilian scientist prepares a hull model *43*
Experimental Model Basin just before its dedication *43*
"USS Maine Blowing Up in Havana Harbor on 15 February 1898" *44*
Rear Admiral William T. Sampson *45*
Sailors prepare a casket for a ceremonial procession *46*
Visitors on a gun barrel *47*
A seaplane catapult launching *49*
President and Mrs. Howard Taft with Argentine visitors *50*
Presidential yacht *Mayflower* *50*
The Great White Fleet underway *51*
President Theodore Roosevelt welcomes the return of the Great White Fleet *51*
Curtiss Pusher plane takes off *52*
Preparing to test a catapult *52*

Ordnance for a Navy Second to None
Gun factory workers processing steel *53*
A tractor mount and 7-inch naval rifle *54*
Pay wagon *55*
Navy Yard Chapter of the American Red Cross *55*
Secretary of the Navy Josephus Daniels *56*
American sailors preparing to emplace a mine *56*
Submarine chasers searching for German U-boats *57*
A 14-inch naval rifle mounted on a railway car *58*
Seaplane model in the EMB wind tunnel *58*

Between the World Wars
    Industrial activity at the gun factory 59
    Navy yard workers honored 60
    Navy yard employee Almira V. Brown 61
    Camp Good Will sponsored by the navy yard 61
    Captain J. J. Raby throws out the ball on opening day 62
    Navy yard baseball team cartoon 62
    United States Navy Band performing at the navy yard 62
    The first aircraft carrier *Langley* (CV 1) 63
    Charles A. Lindberg 63
    Mayflower during a 1912 fleet review 65
    Thomas King, 66
    Flood waters inundate navy yard buildings 66
    President and Mrs. Herbert Hoover with Commander Louis J. Gulliver, CO of *Constitution* 67
    Presidential yacht *Potomac* arrives at the navy yard, 68
    President and Mrs. Roosevelt with King George VI and Queen Elizabeth 68

Ordnance Nerve Center for a Global Conflict
    A skilled worker clears metal cuttings from a gun barrel 69
    Employees work metal into shape 70
    Work in the breech mechanism shop 71
    A 3-inch .50-caliber, rapid-fire, twin-mount antiaircraft gun 71
    World War II sailors test gun range-finders 72
    Sparks and flames leap from cauldrons of molten metal 72
    Sailors take part in navy yard ceremony 73
    Navy yard employees gather in Leutze Park for a band concert 74
    Women shift workers take a break 74

A New Mission for the Navy Yard
    President Harry S. Truman 75
    1947 map 76
    Naval School of Diving and Salvage 77
    Submarine torpedo tubes produced at the gun factory 77
    Welcoming ceremony for Rear Admiral Richard E. Byrd 78
    Navy yard workers operate data processing equipment 78
    Land acquired by the Washington Navy Yard 79

President Dwight D. Eisenhower disembarks *Mayflower* 80
USS *Blandy* (DD 943) 80
Wisconsin (BB 64) fires a 16-inch gun 81
Washingtonians enjoy a "Watergate Concert" 81
Admiral Arleigh Burke, former Chief of Naval Operations 82
Fighting top of USS *Constitution* 82
Guns and missiles on display in Willard Park 83
The Navy Museum 83
A PT boat in President Kennedy's inaugural parade 84
President John F. Kennedy speaking at the navy yard 84
The Navy Art Gallery 85
"Score Another for the Subs" 85
Children at The Navy Museum 86
First Lady Hillary Rodham Clinton, Mayor Anthony Williams, Rear Admiral Arthur N. Langston, Congresswoman Eleanor Holmes Norton, and Director of Naval History William S. Dudley 87
"Queen of the Fleet" 88
Captain William F. McGonagle, Commanding Officer of *Liberty* 88
USS *Liberty* (AGTR 2) after attack by Israeli forces 89
Presidential yacht *Sequoia* 89
Captured 18th-century guns in Leutze Park 90
Civil War-era cannon along Dahlgren Avenue 91
Commodore Dudley W. Knox 92
Assistant Secretary of the Navy James R. Soley 92
The Dudley Knox Center for Naval History 93
The Marine Corps Historical Center 94
Sailors and marines render honors 95
Aerial view of the Washington Navy Yard looking south 96–97
Aerial view of the Washington Navy Yard looking north, 98
Artist's rendering of commercial and government facilities, 99

Unless otherwise specified, photographs are from the collections of the Naval Historical Center.

Foreword

It is a great honor to be the 83rd Commandant of the Naval District of Washington, home of the Navy's oldest and most historic base, the Washington Navy Yard. I especially value the opportunity to serve in this capacity in this, the Navy Yard's bicentennial year.

Throughout its history, the yard has been associated with names like Washington, Jefferson, Lincoln, and Kennedy. Kings and queens have visited the yard; its waterfront has seen many historic moments; and some of our Navy's most senior and most notable officers have called it home. Such legendary ships as USS *Constitution* and USS *Constellation* sailed from its piers, and the 14-inch and 16-inch guns that armed our Navy's battleships during Word Wars I and II were built in its factories.

The Navy Yard's colorful and storied history was originally chronicled in *Round-Shot to Rockets: A History of the Washington Navy Yard and U.S. Naval Gun Factory*, written in 1949 by Taylor Peck to mark the yard's sesquicentennial. His work is indeed a comprehensive history of the first 150 years of the Washington Navy Yard. It details the evolution of the yard from its earliest days and documents the service of such legendary officers as Captain Thomas Tingey, who served as the first Commandant for an astounding twenty-nine years, and Rear Admiral John A. Dahlgren, perhaps the Navy's most renowned ordnance engineer.

The Washington Navy Yard: An Illustrated History, by Edward J. Marolda, complements *Round-Shot to Rockets* and details the proud heritage of the Navy Yard during the last fifty momentous years. Indeed, this year as we celebrate our bicentennial and face the new millennium, the Washington Navy Yard is undergoing significant growth and revitalization as it takes on a new role as a model for cost-effective support to our Navy's shore establishment. At the same time, the Navy's long-standing commitment to and partnership with the community around the Navy Yard is undergoing a rejuvenation and expansion. The stage is thus set for another era of service to our nation by the Navy Yard.

The Navy Yard is extremely fortunate to have within its walls the Naval Historical Center and its Navy Museum, the display ship *Barry*, and a variety of artifacts that attest to the yard's key role in the growth of a strong and capable U.S. Navy. The Naval Historical Center's important contribution to the preservation of our rich history is reflected in this excellent book and for this I offer my personal thanks to that fine institution.

My hope is that after you read about the Washington Navy Yard, you will make it a point to come here as part of your visit to our nation's capital. In so doing, you will be captured, as I have been, by the historical significance of this wonderful installation. But you will also be honoring the hundreds of thousands of Sailors, Marines, and civilians who have given, and continue to give, a great deal of themselves to our country and our Navy at this historic place.

Rear Admiral Christopher E. Weaver, USN
Commandant Naval District Washington

Preface

During much of the 19th and 20th centuries, the Washington Navy Yard was the most recognizable symbol of the United States Navy in the nation's capital. The shipyard built a number of the Navy's first warships and repaired, refitted, and provisioned most of the frigates, sloops, and other combatants of the fledgling naval service. The masts and rigging of USS *Constitution* were a common site on the banks of the Anacostia River. Booming cannon became a routine sound in southeast Washington during the mid-19th century as Commander John A. Dahlgren, "father of American naval ordnance," test-fired new guns for the fleet. The Naval Gun Factory's fire and smoke-belching blast furnaces, foundries, and mills gave birth to many of the fleet's weapons, from small boat howitzers to the enormous 14-inch and 16-inch rifles that armed the naval railway batteries in World War I and the *Iowa*-class battleships in World War II and the Cold War. Rear Admiral David W. Taylor inaugurated a new era in ship development when he used scientific measurements in his Experimental Model Basin to test the properties of prototype hulls. Before and after World War I, the pioneers of naval aviation experimented in the Anacostia and navy yard facilities with various seaplane types, shipboard catapults, and other equipment that would soon revolutionize warfare at sea.

The Washington Navy Yard has been a witness to history—to the evolution of the United States of America from a small republic, whose ships were preyed upon by Barbary corsairs and whose capital was burned by an invading British army, into a nation of enormous political, economic, and military power and global influence. The Civil War that so dramatically altered American society swirled around and through the Washington Navy Yard. American presidents, first ladies, foreign kings and queens, ambassadors from abroad, legendary naval leaders, national heroes and villains, and millions of citizens have all passed through Latrobe Gate during the yard's 200-year existence.

The Washington Navy Yard has also been the workplace for tens of thousands of Americans, a familiar landmark in the District of Columbia, and a valued member of the Washington community. Throughout the 19th and 20th centuries, ship riggers, hull caulkers, iron and bronze smiths, joiners, millwrights, machinists, foundrymen, boilermakers, and tool and die makers; skilled workmen and laborers; naval officers, bluejackets, and marines have earned their livings within the walls of the navy yard. Numerous Americans, white and black, male and female, have spent their entire working lives at the yard building warships, manufacturing guns, testing vessel and aircraft models, training sailors, or administering the needs of American combatants steaming in the distant waters of the world. Navy yard workers, as many as 26,000 men and women at one point in 1944, contributed to the success of U.S. arms in the Spanish-American War, World Wars I and II, the Cold War, and Operation Desert Storm.

Yard workers, most of them residents of the District, Maryland, and Virginia, over the years have helped local authorities extinguish fires, hold back flood waters, rescue victims of natural disasters, and care for needy members of the surrounding neighborhoods. They have helped federal authorities put together national celebrations to mark the end of the country's wars, repair the Capitol and other government buildings, receive the sacred remains of unknown U.S. servicemen from overseas, stage presidential inaugurations, and welcome foreign dignitaries to American soil. Above all, they have loyally served the United States and the U.S. Navy.

This richly illustrated history was written in the bicentennial year to highlight the importance of the Washington Navy Yard and its employees to the nation, the Navy, and the District of Columbia. It touches on the major activities of the facility and on some of the yard's past workers and significant visitors. Much of the text relies on authoritative compilations, published and unpublished, primary documents, and other sources identified in the bibliography that are maintained in the Navy Department Library and the Navy's Operational Archives. We do not suggest that this is the definitive work on the subject or a follow-on to the seminal history by Taylor Peck, *Round-Shot to Rockets: A History of the Washington Navy Yard and U.S. Naval Gun Factory*, published in 1949. Accomplishing that worthwhile goal demands intensive research in the document collections and personal papers relating to the navy yard in the National Archives, Library of Congress, Operational Archives, and numerous other repositories nationwide. A worthy research and writing project beckons some aspiring scholar!

As the momentous 20th century and the millennium draw to a close, the Navy is increasingly concentrating its Washington-area commands and activities at the Washington Navy Yard. The site offers many financial and administrative advantages for the Navy in the lean budgetary environment of the post-Cold War era. Longstanding buildings, unique in their designs and reminders of the navy yard's proud industrial past, are being reconditioned to accommodate the Naval Sea Systems Command and other commands. New government and commercial buildings are rising along the banks of the Anacostia and in the nearby southeast Washington neighborhoods. The Washington Navy Yard is once again abuzz with activity. Hence, in this momentous bicentennial year, the future looks especially bright for the U.S. Navy's oldest and most historic base.

Edward J. Marolda

Acknowledgments

I owe special thanks to Rear Admiral Christopher E. Weaver, Commandant Naval District Washington, for his encouragement and financial support of this publication project and for his energetic efforts highlighting the Washington Navy Yard's significance in American history. Additional support came from John Imparato and Senior Chief Jody Haubry of the admiral's staff, who worked hard to bring about successful celebrations in this momentous year.

It is especially appropriate that professional staff members of the Naval Historical Center, many of whom have worked in the Washington Navy Yard for decades, some for their entire professional careers, have collaborated in the completion of this anniversary history. While many of my colleagues long ago dedicated heart and soul to the study of naval history, they have taken special interest in interpreting the evolution of their workplace. I wish to thank Dr. William S. Dudley, Director of Naval History, who first proposed preparation of this work and then gave unwavering support to its completion; Regina Akers, Christine Hughes, Jack Green, Kim Nielsen, John Reilly, Mark Evans, Dr. Robert J. Schneller, and Susan Scott, learned professionals who authored sidebars and reviewed the manuscript; Charles R. Haberlein, Jack Green, Edwin Finney, and Roy Grossnick, whose knowledge of the Center's photo collections enabled them to suggest and locate unique images; Gale Munro, Lynn Turner, Karen Haubold, and DM1 Erick Marshall Murray, who recommended several exceptional paintings from the Navy Art Collection; Jean Hort, Glenn Helm, Barbara Auman, Tonya Montgomery, and Davis Elliot, who pointed me to rare and valuable published and unpublished manuscripts in the Navy Department Library; Kathy Lloyd and Regina Akers, who suggested key sources in the Operational Archives; Captain Todd Creekman and Commander James Carlton, successive Deputy Directors of Naval History, administrator Jill Harrison, and financial manager Donna Smilardo, who provided encouragement and administrative support. As with every publication project of the Center, Senior Editor Sandra Doyle applied her considerable professional skills to the editing, composition, and design of the work.

Numerous individuals outside the Center eagerly contributed to this history. James Dolph, in addition to carrying out his responsibilities as the historian of the Portsmouth Naval Shipyard, conducted extensive research on the industrial uses of Washington Navy Yard buildings and graciously shared the product of that work with the author; Carolyn Alison, Public Affairs Officer of the Office of the Judge Advocate General, and Roger Williams and William Brack of the Naval Facilities Engineering Command helped enlighten the author about the current restoration of historic navy yard buildings; Jan Herman, historian of the Bureau of Medicine and Surgery, provided much useful information on Washington's earliest naval medical facility; Captain Guillermo Montenegro, Argentine Navy (Ret.), made a special effort to verify the visit by President William Howard Taft to Argentine warship *Sarmiento* when the vessel put in at Washington in 1910, and Captain Hugo Diettrich, Argentine Navy (Ret.), provided a photo of the event; Dr. David Winkler and David Manning of the Naval Historical Foundation facilitated the reproduction of photographs; Lawrence P. Earle and A. Donald Lawrence of the Chesapeake Division, Naval Facilities Engineering Command supplied several of the graphics and aerial photos; Master Chief Musician Jon Youngdahl of the United States Navy Band provided historical information on his long-time Washington Navy Yard tenant activity; Jean Kirk of the Medals and Awards Branch made available unique photographs; and DiAnn Baum, Bill Rawley, and Jeffrey Dorn of the Government Printing Office's Typography and Design shop applied their expertise to the final format and design of the book.

I am grateful to all those individuals who helped in the preparation of this illustrated history, but any misinterpretations or factual errors that appear on the following pages are my own. The views expressed do not necessarily reflect or represent those of the Department of the Navy or any other U.S. Government agency.

Edward J. Marolda

"Captain Thomas Tingey" (Navy Art Collection, artist unknown). Tingey was the first Commandant of the Washington Navy Yard. During the thirty-year tenure of this courageous and focused officer, the yard became one of the Navy's prime shipbuilding and supply centers.

The Early Years

The Washington Navy Yard dates its founding from the earliest days of the republic and the United States Navy. The Continental Navy that fought the American Revolution from 1775 to 1783 was disestablished soon after the conflict. Congress and many Americans concluded that there was no longer a need for naval protection. The citizens of the new United States of America soon learned that they could not prevent European navies or privateers commissioned by Great Britain, France, and the Barbary states of North Africa from seizing American merchant ships on the high seas. So in 1794 Congress appointed six naval captains and approved funds for the construction of *Constitution* and five other fast, 44-gun and 36-gun frigates. The commissioning of *United States* in 1797 marked the rebirth of the American navy.

The following year, Congress established the Department of the Navy and appointed prominent Georgetown merchant Benjamin Stoddert as the first Secretary of the Navy. A dynamic and visionary leader, Stoddert worked hard to build and develop the new U.S. Navy. He hoped to limit shipbuilding costs by establishing naval yards on federal land at several East Coast locations rather than depend on private firms for construction. Knowing that President George Washington had given his approval in 1792 for a 12-acre tract on the Eastern Branch (now the Anacostia River) of the Potomac River, President John Adams approved use of the public land at the site in 1798. Stoddert had affirmed for the President that this site in the new federal capital lay near abundant supplies of timber for building warships and possessed a suitable waterfront from which to launch them. Equally important to the Secretary of the Navy and the nation's founders, who were suspicious of standing military forces, a navy yard at that location just off Capitol Hill could be kept "under the eye of the Government."[1] On 25 February 1799, Congress passed legislation that provided $50,000 for construction of two drydocks and $1 million for the launching of twelve warships. The drydocks and one ship were to be completed in Washington.

The Washington Navy Yard began its illustrious history on 2 October 1799, when public land on the

The Tripoli Monument. This memorial to Stephen Decatur and other heroes of the Barbary Wars was the national capital's first monument. It greeted visitors to the navy yard for thirty-five years at the beginning of the 19th century and now rests on the grounds of the U.S. Naval Academy at Annapolis, Maryland.

Eastern Branch passed into Navy custody. With the establishment of the yard, Benjamin Stoddert, who would soon direct U.S. naval forces during the Quasi-War with France, put the Navy on a firm footing as the young, vibrant nation confidently entered the new century.

To develop the Washington Navy Yard and build its first warship, Stoddert selected Captain Thomas Tingey, a short, stout naval officer who had already made a name for himself. The English-born Tingey had served as an officer in the Royal Navy, before joining the U.S. Navy in 1798. He erased any doubts about his loyalty to the new republic when he refused to let British officers search his naval vessel for crewmen who might have been the king's subjects. He had assured his shipmates that he would die before surrendering any of them to the British.

Arriving in Washington in February 1800, Tingey inspected the site of the future navy yard and suggested the need for a contingent of marines and a high fence around the site to discourage pilferage of lumber and other building materials. The Navy Department acted on his suggestion. From that year to the present, the U.S. Marine Corps has continuously posted a guard at the Washington Navy Yard. Tingey also rented nearby a solid, two-story house with a large kitchen, stable, carriage house, and a well. When he moved in during the spring of 1800, Tingey became the first naval officer to make a permanent home in the District of Columbia.

Following the short Quasi-War with France and Thomas Jefferson's presidential inauguration in 1801, the Navy decommissioned a number of its older warships and placed them in reserve or "in ordinary" at the Washington Navy Yard and its other naval facilities. Equipment from the ships and huge piles of live oak and other timber used for constructing sturdy warships were kept out of the elements in a new, covered building on the north bank of the Eastern Branch. That fall, Tingey oversaw the completion of an officer's residence, constructed under contract with the Lovering and Dyer Company of Georgetown. During this same period, a small "naval hospital" was established in a

"Benjamin Stoddert" by E. F. Andrews (Navy Art Collection). Stoddert, the first Secretary of the Navy, pressed for the establishment and early development of the Washington Navy Yard.

building outside the yard near the main gate. The facility, the only one of its kind in the Washington area, provided medical care to navy yard personnel for the next three decades.[2]

President Jefferson, whose home at Monticello near Charlottesville, Virginia, bears witness to his genius for inventive architecture and design, became excited about the prospect of building an 800-foot-long, 175-foot-wide drydock at the Washington Navy Yard. He suggested that with a system of locks Navy frigates (virtually the entire fleet in 1802) could be raised out of the water and kept in a covered structure until they were needed for operations. He reasoned that this facility would significantly reduce the funds that Congress, which continued to oppose maintaining a large navy, would have to appropriate for ship upkeep. To design the structure, he enlisted Benjamin H. Latrobe, a brilliant architect and like himself, a Renaissance man. Latrobe agreed that such a drydock was feasible. Congressmen from the western states and others opposed to even a small navy, however, voted down the appropriations bill.

Latrobe stayed on in the capital city as Jefferson's Surveyor of Public Buildings and after 1804 as Engineer of the Navy Department. In the latter capacity he developed an overall design for the Washington Navy Yard that the Navy accepted the following year. The main gate, which has come to be known as the Latrobe Gate, was built from his plan and is one of the few original structures in the yard. During Latrobe's tenure with the Navy Department in the first decade of the 19th century, the navy yard witnessed construction of the Commandant's House, the main gate and guardhouse, offices, and shops for joiners, blacksmiths, plumbers, armorers, riggers, and painters. Also established were a foundry and rolling mill. A rudimentary steam engine powered many of the yard's machines—when it worked, which it often did not.

Neither Jefferson nor Congress wanted to incur the expense of maintaining an operating fleet, so the Navy's frigates were laid up. By late 1804, the 44-gun frigates *Constitution*, *United States*, and *President* and the 36-gun frigates *Congress*, *Constellation*, and *New York* had been placed in ordinary at the Washington Navy Yard. As an alternative, Jefferson pushed the construction of small, one- or two-gun gunboats, which he considered adequate to defend American ports. The Washington Navy Yard built the prototype gunboat and many of the several hundred follow-on models. The Navy found them unseaworthy and poor combat vessels and by 1815 had taken them out of service.

As they had at the end of the 18th century, the Barbary corsairs and European navies continued to interfere with U.S. oceangoing trade, even in American coastal waters. On one occasion, Jefferson warned the captain of supply ship *Huntress*, then at the navy yard, to watch out for privateers when he put to sea. Despite this warning, the master lost his ship and its cargo to a Spanish privateer only three days out of port. To defend American merchantmen

Tingey House or Quarters A, the long-time residence of Washington Navy Yard commandants and after 1978 the official home of the Chief of Naval Operations.

from these oceangoing predators, Congress funded the building of several sloops-of-war. When completed in 1806, *Wasp* became the first ship built in the Washington Navy Yard.

In the decade before the War of 1812, the navy yard employed several hundred workmen, many of whom lived in a nearby settlement called Navy Yard Hill. This habitation boasted a public market, village green, churches, wells, and springs. Fields of grass or grain, fruit orchards, and shade trees surrounded individual homes. Slaves also toiled at the navy yard.

"USS Chesapeake" by F. Muller (Navy Art Collection). In 1807, HMS Leopard fired on the American warship, killing and wounding twenty-four men. The British removed four sailors from her, claiming they were subjects of their king. The incident, a major irritant to U.S.-British relations, was a factor in the outbreak of the War of 1812.

MOSQUITO SQUADRON.

"Mosquito Squadron" by Fred S. Cozzens (Navy Art Collection). The navy yard built a number of these gunboats during the Jefferson administration.

Their accommodations can only be surmised, but they were undoubtedly much more austere than the homes of Navy Yard Hill.

In 1807 frigate *Chesapeake*, after fitting out at the navy yard, sailed to Norfolk, Virginia, then out to sea where the British warship HMS *Leopard* fired on her, compelling the American captain to submit. Royal Navy officers boarded the U.S. naval vessel and removed several crewmen they believed to be British subjects. Involved in a life and death struggle with Napoleon's France, the British wanted every seaman they could get for their navy, even those men crewing American vessels. United States citizens were outraged by the high-handed British action and there was talk of war. Diplomacy averted a conflict, but many people remembered the slight to national honor when Anglo-American relations again soured in 1812.

As the nation edged closer to war with Great Britain, Captain Tingey oversaw the fitting out of warships at the yard. Not all of his customers were satisfied with the service. On one occasion, Captain Stephen Decatur, by now a national hero for his exploits against the Tripolitan corsairs, complained that the rigging installed on his ship, *United States*, was inadequate. Captain William Bainbridge, a combat officer almost as famous as Decatur, made a similar complaint. Tingey explained to Secretary of the Navy Paul Hamilton that he considered the materials good enough and that some officers were too demanding. Personal judgment was key to the issue, since the Navy then had no standard specifications for masting and sparring a ship, the quantity of stores to be carried, or even arming a warship.

Privateers commissioned by Great Britain, France, Spain, and the Barbary states of North Africa and pirates continued to prey on American merchant

The Navy Yard's Oldest Buildings

The creator of the gateway to the Washington Navy Yard envisioned portals befitting a place of labor. Benjamin Henry Latrobe, President Thomas Jefferson's Surveyor of the Public Buildings, designed the Latrobe Gate in 1805 "with a view to the greatest economy that is compatible with permanence, and with an appearance worthy of the situation." Durable but inexpensive freestone composed the one-story arch whose simple architrave and frieze without a pediment echoed a kinship with Greek Revival design. The north facade comprised two Greek Doric columns set ten feet apart and two flanking pylons. An eagle grasping an anchor in its talon originally crowned the gate and a rectangular panel with an anchor design carved in bas-relief was inset in each pylon. A semicircular arch spanned a 20-foot roadway on the southern façade and a double colonnade of Doric columns connected the two sides. A low-pitched hipped roof covered the structure and the one-story brick guard lodges housed the marine sentries. Completed in 1806, the Latrobe Gate graces the Navy's first navy yard, and its Greek character foreshadowed the coming of a new movement in American architectural thinking that emphasized the affinity between the Greek democratic state and the American republic.

Quarters B or Second Officer's Quarters, the oldest structure at the Washington Navy Yard. Early in the 19th century, local builders Lovering and Dyer created this residence for officers who served as second in command at the naval facility. The building later served as a barrack for the U.S. Marine guard force and then the residence of the Commandant, Naval District Washington.

Alterations in 1880 and 1881 added two stories across the gate and three stories on either side of it, leaving most of the old gate intact, except the eagle, the old roof, and a parapet wall. A Victorian melange of classical and Italianate motifs characterized this new design.

Quarters A is commonly called Tingey House after the yard's first commandant, Thomas Tingey, who lived in the house from 1812 until his death in 1829. Situated on the crest of a hill adjacent to the north wall and just east of the Latrobe Gate, Tingey House, which dates from 1804, is a two-story structure built with Flemish bond brick. A late Georgian design with a symmetrical five-bay front marks the original building, but extensive renovations, retrimmings, and the lengthening of windows in the mid-19th century have made it substantially Victorian. The many additions to the exterior of the house have covered up but not destroyed the original fabric. In the face of the imminent British occupation in August 1814, Thomas Tingey set the yard on fire but Quarters A survived. The building's central location in the yard gave its occupants a sweeping view of the workshops and laborers in the gently sloping vista below the house. Every commandant of the yard resided in Quarters A from 1812 until 1978 when it became the home of the Chief of Naval Operations.

Lack of substantive documentation clouds the origin of Quarters B. One unsubstantiated view holds that it was originally a farmhouse that stood on the yard when the Navy acquired the property in 1799. Another account asserts that Secretary of the Navy Robert T. Smith ordered Commandant Thomas Tingey in October 1801 "to contract with Mssrs. Lovering and Dyer for the building of a house to accommodate the officer of Marines and the Superintendent of the Navy Yard." According to Navy records a contract for a house was let to this firm and in all probability it was constructed in late 1801 or early 1802. The latter version is the more reliable as the house does not appear on the November 1801 enumeration of houses in Washington. Tingey, however, did not live there but chose to rent a house on 11th and G Streets, near the yard. Quarters B was evidently used as an officer's residence and offices during its early years. This structure survived the British invasion of Washington in August 1814, but residents from the neighborhood looted the house when the U.S. Navy abandoned the yard temporarily. Quarters B was originally a simple two-and-one-half-story, gabled-roof, Federal-style brick house approximately 36 feet long and 24 feet deep. The house was enlarged twice to its present length of 76 feet in the years before 1868. Today it houses the Commandant Naval District Washington.

Designed originally as an office for the officers of the Washington Navy Yard, Building 1 or the Commandant's Office was built between 1837 and 1838. Latrobe Gate is the ceremonial entrance to the Washington Navy Yard from Eighth Street and Building 1 forms the southern terminus on this axis. Exterior modifications to the building have not obscured its original scale or mass. The structure is a freestanding, rectangular, two-story brick building laid in a Flemish bond pattern. A low, bellcast hipped roof covers the main block, and frame porches surround the exterior. Renovations in 1873 and in 1895–1896 changed the style of porch columns and railings. In the later renovation the north façade gained a projecting, one-story entrance porch with a gabled roof, and the south façade, a projecting two-story porch with stairs. Besides being a major focal point of the yard, Building 1 served as the administrative center for its commandants. John Dahlgren, perhaps the yard's most illustrious figure, slept and ate in a room outside his office in 1861, having proffered the officers of the 71st New York Infantry Regiment accommodations in Tingey House. President Abraham Lincoln and his cabinet members frequently visited Dahlgren in Building 1 during the Civil War. In the mid-20th century, Building 1 served as living quarters; later it was a Public Affairs Office. Since 1993, historians of the Naval Historical Center have occupied the Navy's oldest active office building.

The National Register of Historic Places has included Latrobe Gate, Quarters A (Tingey House), Quarters B, and Building 1 on its listings since 1973.

Christine F. Hughes

The old Commandant's Office (Building 1).

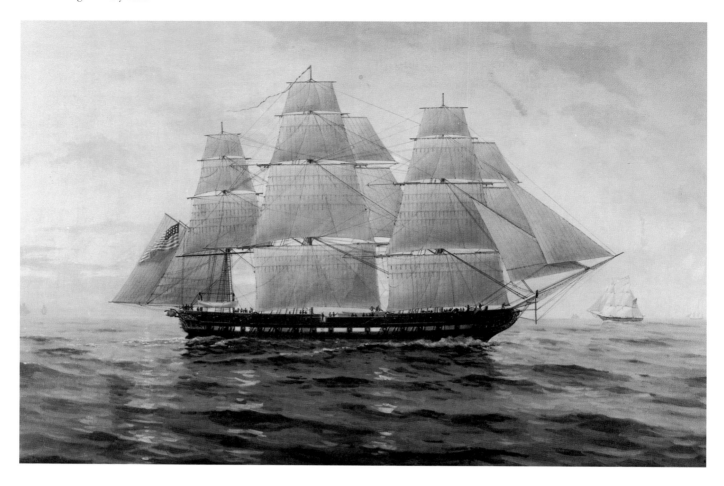

shipping during the early years of the 19th century, but opponents of the Navy in Congress refused to fund new warship construction. Many legislators from the southern and western states, which were not dependent on overseas trade, could see no benefit to a larger Navy.

Even though several of the nation's other navy yards fell on hard times for lack of adequate funding, the Washington Navy Yard did get support for repairing and refitting the Navy's warships. From 1810 to 1812, the yard shops replaced rotten timbers and weathered rigging, caulked and coppered the hulls, and painted sections of brigs *Vixen* and *Hornet*, schooner *Enterprise*, frigate *Adams*, and sloop *Wasp*. Tingey observed in his correspondence that the yard had had to replace only one rotten timber in *Wasp*, which was built in his yard, while the privately built *Hornet* had to be completely rebuilt to replace many rotten timbers. The yard also equipped ten gunboats with new masts, sails, and cannon and added heavier ordnance, stronger masts and spars, and new rigging to frigates *Congress*, *Constellation*, and the soon-to-be famous *Constitution*.

"USS Constellation" by John W. Schmidt (Navy Art Collection). The ship was one of the first frigates built for the U.S. Navy and a common site at the navy yard in the early 19th century.

The War of 1812

The United States declared war against Great Britain in 1812, and at the beginning of 1813 Congress authorized construction of four 74-gun warships and six 44-gun frigates. The Washington Navy Yard almost immediately began work on frigate *Columbia* and sloop-of-war *Argus*. Preoccupied with military operations on the frontier with Canada, the Madison administration failed to improve Washington's defenses. As a stopgap, in 1813 navy yard officers and seamen manned three small scows armed with a few light pieces of ordnance and took station in the Potomac River off Greenleaf's Point. Tingey emplaced and fortified a battery of seven guns on the point (now part of Fort Leslie J. McNair).

"Joshua Barney" by J. Gross (Roosevelt Library). The commodore made a heroic stand with sailors and marines at the Battle of Bladensburg.

When it appeared that the British were headed for Baltimore, the Navy ordered all but a few officers and men at the yard to join Commodore Joshua Barney and his Chesapeake defense flotilla. Service in the local militia companies also took away many of the yard's civilian workers.

The vulnerability of the nation's capital to attack became clear when a British fleet commanded by Rear Admiral George Cockburn and carrying troops under Major General Robert Ross sortied from Bermuda, entered the Chesapeake, and moved north to the mouth of the Patuxent River in August 1814. When the alarm was raised in Washington, the government called out local militia units under Brigadier General William H. Winder. His motley force of regulars and ill-trained militiamen boldly marched across the navy yard bridge on the Eastern Branch and took up positions at Bladensburg, Maryland, some twelve miles to the northeast.

At the same time, Commodore Barney and the sailors and marines of his small defensive flotilla, trapped in the upper reaches of the Patuxent, burned their vessels and headed for the navy yard bridge, which they initially were ordered to defend. When General Winder realized on the 24th that the British troops were not headed for the bridge, he dispatched Barney and his men to Bladensburg. When the commodore's group arrived, many American regulars and militiamen had begun to flee from the battle being fought there. Barney deployed his sailors and marines and their guns to halt the enemy advance, and despite repeated assaults the British regulars could not budge the naval force. Finally, when the enemy routed an American unit on one flank and wounded Barney with their fire, the sailors and marines retired toward Washington.

Meanwhile, Captain Tingey prepared to carry out the orders of the Secretary of the Navy to burn the Washington Navy Yard, rightly characterized by one writer as the "pride of the U.S. Navy," lest it be captured by the enemy.[3] He observed that the navy yard was "historic acreage...where the fledgling navy's warships were built." He added, "Tingey and his staff were keenly aware of the navy yard's place in the tiny capital. It was not only the workplace for hundreds of Washingtonians, but was also one of the most familiar landmarks in their hometown."[4] The gravity of Tingey's mission was driven home in the late afternoon when Winder's men ignited explosives that completely destroyed the bridge over the Eastern Branch.

The commandant dispatched his chief clerk, Mordecai Booth, on a scouting mission to determine if British troops were advancing on the navy yard. When the enemy made clear their approach by twice firing on Booth, he hurried back to report to Tingey. Even then the captain hesitated. He did not want the enemy to capture the valuable navy yard and the ships and stores there, but he also did not want to start a fire that might spread to neighboring houses. Tingey delayed giving the order until after dusk, when the wind died down. Then, around 8:20 he ordered fires started in storehouses, offices, mechanical shops, and the sawmill, all of which were crammed full of valuable materials, machinery, equipment, and papers. And, "like possessed pyromaniacs, the government employees set fire to [frigate *Columbia* and sloop-of-war *Argus*, then under construction] and watched pitifully as the flames devoured months of fine workmanship in the brilliant nocturnal pyre."[5] As Tingey and his chief subordinates departed the navy yard in a small vessel, they watched flames consume the facility. Neither Tingey nor the British, however, torched the Commandant's House or the second officer's quarters.

"United States Ship-of-the-Line Columbus" (Navy Art Collection, artist unknown). Built at the navy yard, she served the nation for much of the early 19th century.

The next day, enemy troops briefly entered the yard and burned the buildings that had not been set on fire by the Americans. Surprisingly, the Americans and the British failed to destroy schooner *Lynx*, which was close to completion; two gunboats; and stores of powder and other materials. As soon as the British departed, the local populace began plundering not only the untouched stores but also the two residences. As one historian observed, "not a movable object from cellar to garret was left, and even the fixtures and locks on the doors were taken."[6] Soon afterward, Tingey strongly recommended erecting a 10-foot-high brick wall around the yard to prevent a recurrence of this activity.

Supporting the New Navy

In the years after the War of 1812, Congress appropriated funds to build new structures at the navy yard according to the previous street layout, but this time made of brick. The Navy also considered different uses for the yard. On the recommendation of the new Secretary of the Navy, Benjamin Crowninshield, Congress established a Board of Navy Commissioners, composed of three senior Navy captains, to advise him on the administration of the naval service. The first board recommended that the number of navy yards be reduced to three and include the Washington Navy Yard. They also observed that because of the difficulty that the larger vessels experienced navigating the Potomac River and its Eastern Branch and because of Washington's distance from the sea, most of the fleet should operate from and be repaired at sites on the ocean. Yet, they found that the Washington site remained indispensable as a shipyard and industrial facility because of its skilled workforce and unique shops.

In that regard, shops were established for manufacturing the Navy's anchors and chain cables. With a new steam engine, the navy yard could fabricate anchors more economically than private firms and ensure product uniformity for the Navy. The establishment of a factory for making chain cable made the yard the sole source for the naval service. In 1827, the Board of Navy Commissioners added official weight to the navy yard's status as the manufacturer for the Navy when they stipulated that many items of ships' equipment were to be provided only by the yard.

The Navy's noteworthy performance during the War of 1812, in marked contrast to that of the Army,

"USS United States versus HMS Macedonian" by Arthur N. Disney Sr. (Navy Art Collection). U.S. Navy sailors gained a reputation for professional skill and valor in battle during the War of 1812.

persuaded the American people and their representatives in Congress to build a postwar fleet worthy of the new, proud nation. The legislature provided for the building of nine ships of the line and twelve 44-gun frigates at navy yards, rather than at private yards whose products were often considered unsatisfactory. In 1819, the Washington Navy Yard launched the 74-gun *Columbus,* which served the Navy ably until her deliberate destruction to prevent capture during the Civil War.

Completion of the frigate *Potomac* coincided with Commodore John Rodgers' development of a new aid to construction and repair. He considered it possible for yard workers to haul a ship out of the water on an inclined plane and into a roofed-over building so work could proceed regardless of wind, rain, and snow. The commodore demonstrated the feasibility of his project before President James Monroe, members of Congress, foreign dignitaries, and the public in 1822. Using ropes, 140 men hauled *Potomac* out of the water and along a wooden incline, the nation's first marine railway. So impressed were his visitors that in 1823 Congress authorized construction of a marine railway to be powered by men, animals, or mechanical winches and a large, covered shiphouse. The completed building became a recognized feature of the Washington Navy Yard for many years. The first ship worked on in the shiphouse was *Potomac,* which took part in the siege of Vera Cruz during the Mexican War and operated with the Union blockading forces in the Civil War.

From the beginning, the navy yard was closely involved with the activities of the federal government in the nation's capital and with the life of the Washington, D.C. community. For instance, skilled yard workers repaired a chandelier that fell from the ceiling of the House of Representatives. On several occasions during this early period,

Above. *"President John Quincy Adams"* by Asher B. Durand (Navy Art Collection).

Left. *"USS Brandywine"* by N. Cammilliari (Navy Art Collection). The ship carried Lafayette home to France in 1825.

navy yard men left their workstations to help battle fires in Alexandria and Georgetown.

Throughout its history, the navy yard was host to distinguished American and foreign visitors. President John Quincy Adams stood on the deck of frigate *Brandywine* as she entered the Eastern Branch. Yard workers cheered heartily, not only because of the President's presence but because they were given the rest of the day off to mark the occasion. Originally called *Susquehanna*, the ship had been renamed to honor French General Marquis de Lafayette, a hero of the American Revolution who showed exceptional leadership during the Battle of Brandywine. The general and his party made

"The Marquis de Lafayette" (Pierpont Morgan Library, artist unknown). A French hero of the American Revolution.

M.^r le Marquis DE LA FAYETTE
Commandant General de la Garde Nationale et Citoyenne de la Ville de Paris

a triumphal procession through the streets of Washington and enjoyed a full day of ceremonies at the White House, Capitol, and navy yard on 15 October 1824. Mrs. Tingey hosted a tea for the distinguished French general. The following September, to the cheers of many Americans lining the Potomac, the general boarded *Brandywine*, then anchored in the river. As the ship prepared for the transatlantic journey to France, Lafayette bade farewell to the country whose freedom he helped win.

The yard entered a new era on 23 February 1829 when Thomas Tingey, Commandant of the Washington Navy Yard for almost three decades, died in his quarters. That April, Commodore Isaac Hull, captain of *Constitution* during her famous fight with HMS *Guerriere* in the War of 1812, became the next commandant. The New Englander immediately directed that excess stores, equipment, and building materials be sold or shipped to other yards, shops be cleaned and reorganized, and the yard be made generally shipshape. Change is not often well received, but Hull's personnel actions caused more than the usual irritation. He cut the pay of some workmen and reduced the number of employees in several shops. These and other actions caused friction between Hull and his workers.

Matters came to a head in 1835 when Hull, frustrated by the loss of tools and stores to theft, forbade workers' bringing their food containers into the yard. Even though one workman was discovered secreting government materials out of the yard in his food basket, the others were outraged that their integrity was questioned. Around 150 men walked off the job and marched to the Navy Department building near the White House. Cooler heads finally prevailed and after compromises were made on both sides, the men returned to work. In the fall of that difficult year, Hull asked for and was given permission to resign as commandant. In the years after his resignation, several officers served as commandant; two died during office in the Commandant's House.

"USS Water Witch" by R. G. Skerrett (Navy Art Collection). The vessel was one of the Navy's first paddle wheel steamers.

During this period, the Navy investigated the use of steam propulsion for its warships and by 1842 the navy yard had become the prime site for this work. Manufactured in the yard were the boiler and machinery for *Water Witch*, a small iron paddle-wheel steamer that served the Navy during the Mexican War. That war confirmed the benefits of steam over sail power for warships because steam-powered combatants could operate along coasts and in rivers where there was often no wind to fill sails.

Partly to improve the Navy's administrative efficiency, in 1842 the Secretary of the Navy disestablished the old Board of Navy Commissioners and established six bureaus. The Chief of the Bureau of Yards and Docks was responsible for the Washington Navy Yard, but because the yard handled a number of functions for the service, other bureaus also exercised some authority at the facility. Disagreements between the bureaus often made life difficult for the commandant.

As one of the earliest naval engineering facilities, the yard became the first to produce complete steam propulsion machinery for a warship in 1843, with the commissioning of steam gunboat *Union*. Between 1843 and 1861, the navy yard was the only government facility manufacturing steam plants for naval vessels, and did so for four of them.[7] In 1854, the navy yard constructed the 243-foot-long steam frigate *Minnesota*, which displaced 4,800 tons and was then one of the world's largest naval vessels.

The Father of Naval Ordnance

The development of ordnance for the Navy, for which the Washington Navy Yard became perhaps most famous, was inspired by a tragedy in the Potomac near the nation's capital. On 28 February 1844, President John Tyler and 500 other government officials, foreign dignitaries, and their family members were embarked on *Princeton* for the test firing of a new naval gun, called the "Peacemaker." After several firings, the gun suddenly exploded, killing the Secretary of State, Secretary of the Navy, and a number of other passengers and crewmen. The Navy clearly needed a better system for determining the worthiness of ordnance for its warships.

In 1845, under the direction of Commodore Lewis Warrington, the first Chief of the Bureau of Yards and Docks, the yard established a facility for the study and production of fuses, rockets, and mine-torpedo devices. While testing underwater explosive devices in the Eastern Branch, renowned inventor Samuel Colt sank a vessel whose hulk hindered navigation in the waterway until just before the Civil War.

Few naval officers were as closely associated with the Washington Navy Yard as John A. Dahlgren. Many have characterized him as the "father of American naval ordnance." He became "synonymous with naval armament during the Civil War era" and "an immortal figure in the pantheon of American naval heroes."[8] Dahlgren's genius for ordnance design and production was evident soon after his assignment to the navy yard in 1847 to handle the development of rockets. He was also charged with inspecting rocket fuses, shells, locks, and powder tanks. By the end of the year he was made responsible for handling all ordnance matters at the yard, independently of the commandant.

The navy yard was already the Navy's "principal industrial center" and "foremost metalworking center," whose rolling mill, tilt hammer, foundry, and

"Awful Explosion of the 'Peacemaker'" by N. Currier (Navy Art Collection).

Above. *Interior view of the experimental battery built and operated by Dahlgren on the waterfront of the navy yard. Much of the ordnance operated by the U.S. Navy during the Civil War was first tested and "proofed" at the unique facility.*

Left. *A naval officer rests his arm on the distinctive "soda-water bottle" shaped 9-inch Dahlgren gun, one of the most powerful and reliable weapons in the U.S. Navy's Civil War arsenal.*

blacksmith, anchor-making, rigger, and painter shops manufactured anchors, small arms, chain, and other materiel.⁹

A few years earlier the Navy had decided to standardize its large-caliber guns and needed a system to determine how well prototypes would perform before proceeding with production. Dahlgren had the answer. He built a firing range on the yard's waterfront. Guns being tested fired their rounds out over the Eastern Branch. By recording the splash of rounds, Dahlgren could

Commodore Charles Morris, a hero of the War of 1812, but to Dahlgren an "old fogy."

determine the power and range properties of each type of gun. Like many of his successors, President James K. Polk visited the navy yard to witness the test firing of naval cannon at Dahlgren's "experimental battery."¹⁰

Despite Dahlgren's attention to safety, ordnance testing was dangerous business. Michael Shiner, a slave who worked in the yard from 1813 to 1865, noted in a later account that there were many industrial mishaps and accidental explosions in this period that killed or injured workers.¹¹ In 1849 a gun being tested at the yard's experimental battery exploded. Jagged pieces of iron from the weapon killed the gunner and narrowly missed Dahlgren. This incident reinforced his belief that he needed a different approach to gun development if he were to build safe, large-caliber guns for the Navy. The result was a family of guns, soon called Dahlgren Guns, that had stronger and thicker metal around their breeches, making them look to some people like "soda-water bottles."

Dahlgren's unconventional ideas about ordnance design met strong opposition from some quarters. Commodore Charles Morris, Chief of the Bureau of Ordnance and Hydrography in the early 1850s, delayed or opposed many of Dahlgren's requests for material or other support. Dahlgren considered Morris an "old fogy."¹² Dahlgren overcame most opposition by cultivating friendships with prominent government and congressional figures and other naval officers and persuading them that his ideas were sound. Shortly after Morris died in 1856, the Navy Department implemented Dahlgren's program by arming the new warship *Merrimack* with a battery of his 9-inch guns. He successfully tested the weapon during the ship's cruise from Norfolk to Annapolis. Dahlgren 9-inch and 11-inch guns became mainstays of the fleet and passed the even more demanding test of combat during the four years of the Civil War. In addition to cannon for major warships, Dahlgren designed and had cast in the navy yard brass furnace a 225-pound boat howitzer for use by naval landing parties.

As Dahlgren's ordnance ideas took hold, the Navy built new facilities at the yard to handle the entire design, testing, and manufacturing process. A cannon foundry, mechanical finishing shop, and gun-carriage shops soon joined the existing facilities at the Navy's industrial plant. Dahlgren also worked to create a skilled, dedicated, and professional workforce at the yard. He raised the pay for some civilian workers, established billet descriptions and work standards, and provided relevant training. "By attending to the needs of his men, Dahlgren produced a cadre of loyal ordnance specialists."¹³

The Father of American Naval Ordnance

John Adolphus Bernard Dahlgren (1809–1870) determined the direction of weapons development in the United States Navy for more than two decades (1847–1870). His achievements earned him the title "father of American naval ordnance."

Dahlgren joined the Navy in 1826 and received training in advanced mathematics, scientific theory, and the use of precision instruments while serving on the U.S. Coast Survey (1834–1837). Ordered to ordnance duty at the Washington Navy Yard in 1847, Dahlgren began his first task there—developing rockets for use on board warships. He did the job so well that the Chief of the Bureau of Ordnance and Hydrography assigned him ever increasing responsibilities, empowered him to expand the yard's facilities at his discretion, and arranged for him to remain on ordnance duty indefinitely. Dahlgren served in Washington for the next fifteen years.

In the early 1850s he launched the Ordnance Establishment, the first sustained weapons R&D organization and program in American naval history. Its work fell into four broad categories: manufacturing ordnance and equipment, inspecting ordnance produced at private foundries, testing ordnance and inventions, and research and development. The facilities that Dahlgren set up at the Washington Navy Yard included a firing range along the Anacostia River dubbed the "experimental battery," foundries, machine shops, and expanded office spaces. These facilities became the seed of the Naval Gun Factory.

Dahlgren's R&D efforts yielded integrated systems of shipboard armament, featuring light bronze boat guns, heavy smoothbore shell guns, and heavy rifled cannon. The boat guns were revered throughout the Navy and admired around the world. The heavy smoothbores, characterized by their peculiar "soda-water bottle" shape, were his most famous invention. Their design stemmed from Dahlgren's scientific research in ballistics and metallurgy. To ensure their reliability, he had them manufactured under the most innovative and comprehensive quality control program seen in the Navy to that time. The program included detailed specifications, production monitoring, and a rigorous system of proof testing. Dahlgren also instituted a gunnery training program to teach naval officers and enlisted men how to use the new weapons. When first introduced to the fleet in the mid-1850s, Dahlgren guns were the most powerful and reliable naval cannon in the world. They remained the Navy's standard armament until replaced by steel rifled cannon in the 1880s. Dahlgren's only failure was his heavy rifled ordnance.

In the process of inventing and developing cannon, Dahlgren articulated an unusually comprehensive conception of naval ordnance, including its role in naval strategy and tactics and national policy. An advocate of steam propulsion, he devised doctrine for what he considered the optimum balance of steam, sail, and ordnance in naval vessels. And he improved ordnance equipment of all kinds, including locks, sights, carriages, the rope used for breeching tackle—even the preservative coating used to protect gun barrels from rust.

Promoted to commander in 1855, captain in 1862, and rear admiral in 1863, Dahlgren became Commandant of the Washington Navy Yard in 1861 and Chief of the Bureau of Ordnance in 1862. He spent most of the first two years of the Civil War meeting the Navy's vastly increased demands for ordnance, for which he received a vote of thanks from Congress. During that same period Abraham Lincoln drove his carriage down to the yard almost weekly. He often did so just to have coffee, cigars, and a chat with Dahlgren, who had become the President's favorite naval officer, principal uniformed advisor on naval affairs, and friend.

During the Civil War Dahlgren's guns were mounted on almost every Union ship and many Confederate naval vessels, including USS *Monitor* and CSS *Virginia*, which fought history's first duel between ironclad warships, and USS *Kearsarge*, which sank CSS *Alabama*, the most successful commerce raider of all time. Not one of Dahlgren's 9- or 11-inch guns—the principal heavy calibers—failed prematurely in combat.

With help from Lincoln, Dahlgren left Washington in July 1863 and took command of the South Atlantic Blockading Squadron. For the next two years, he led naval forces besieging Charleston, South Carolina, the Union navy's most frustrating campaign. After the war, he commanded the South Pacific Squadron, the Bureau of Ordnance, and once again the Washington Navy Yard, where he died in 1870.

Dahlgren was an intelligent, hard-working individual who possessed enormous persistence, drive, and self discipline. He gave American sailors something they never had before: an unshakable faith in their guns, a faith justified by the existence of an organization founded on the idea that systematic research and development could produce safer and more powerful weapons. That idea was Dahlgren's most important contribution to the American naval tradition.

Robert J. Schneller Jr.

A view of the navy yard waterfront captured in 1866 by famed photographer Mathew Brady. The building in the left foreground houses the experimental battery.

The first Japanese delegation to visit the United States poses with U.S. naval officers soon after the foreign group's arrival in May 1860. Pictured in addition to Japanese 1st Ambassador Shinmi Buzen (seated 4th from the left) are two men who would figure prominently in American naval history, Lieutenant David Dixon Porter (standing 4th from the left) and Captain Franklin Buchanan (standing 7th from the right).

The ordnance facilities at the yard cast the guns and manufactured their ammunition, and Dahlgren tested them in his experimental battery on the Eastern Branch. In addition to the heavy Dahlgren shell gun and the smaller boat gun, Dahlgren considered development of the "Ordnance Establishment [that]... he created to test, improve, and manufacture naval ordnance" as his most important accomplishment.[14] Ordnance experts from Great Britain, France, Russia, and Sweden visited the Washington Navy Yard to learn more about Dahlgren's designs and investigate the facility that armed the American fleet.

As the nation edged closer to civil conflict in the spring of 1860, the navy yard served to support U.S. foreign policy. On 14 May, the steamer *Philadelphia* brought to Washington the first Japanese delegation to the United States, headed by 1st Ambassador Shinmi Buzen. Captain Franklin Buchanan, Commandant of the Navy Yard and a participant in Commodore Matthew C. Perry's historic 1854 visit to Japan, advanced between two rows of marines at present arms to greet the Japanese dignitary and his party. The foreign delegation, carrying with them in an ornate box the treaty Perry helped conclude, then boarded carriages for the short ride through Latrobe Gate to Willard's Hotel on Pennsylvania Avenue. A little more than a week later, the Japanese party returned to inspect the anchor, boiler, engine, and ordnance shops at the yard. The visitors paid close attention on the tour, for Japan was then in the throes of its own civil conflict, the Meiji Restoration, during which the Japanese incorporated many Western military and industrial ideas.

Civil War Naval Arsenal and Operating Base

As war clouds gathered in early 1861, Captain Buchanan (even though he harbored Southern sympathies) and Commander Dahlgren concentrated cannon, small arms, and ammunition in the yard's main building, "turning it into a fortress," and prepared to blow up the munitions rather than allow their seizure by pro-southern mobs.[15] The threat was real, for southern sympathies were strong in the Maryland and Virginia regions surrounding Washington. John Brown's short-lived capture of the federal arsenal at Harpers Ferry, West Virginia, and the loss of Navy property in Pensacola, Florida, were fresh in the minds of these officers. Only 1,000 troops and 1,500 volunteers guarded the nation's capital. The naval officers drew up defensive plans for the yard's loyal sailors and marines.

The inauguration of Abraham Lincoln as President of the United States on 4 March 1861 triggered South Carolina's attack and capture of federal Fort Sumter in Charleston and secession from the Union of an increasing number of southern states. After Virginia seceded and Maryland prepared to take the same step, Captain Buchanan and every southern-born officer except one at the navy yard resigned his commission in the U.S. Navy. At the President's suggestion, on 22 April 1861, Secretary of the Navy Gideon Welles appointed Commander Dahlgren Commandant of the Navy Yard.[16]

The new commandant had to act fast to defend not only the navy yard but Washington itself, because in the first chaotic months of the Civil War there were few Union troops and even fewer warships available

A U.S. Marine detachment, part of the navy yard garrison, stands at ease during the Civil War. Marines have continuously manned the guard post at Latrobe Gate since the establishment of the naval facility.

for defense. Dahlgren immediately armed civilian steamer *Mount Vernon* and assigned some of his sailors and marines to the ship. On the 23rd, *Mount Vernon* joined five other warships patrolling the lower reaches of the river. He also positioned guns to defend Latrobe Gate, barricaded and fortified the other entrances, and lined the waterfront with cannon to oppose a waterborne assault on the yard or attack against the bridge across the Eastern Branch. The defenders of the navy yard included thirty-seven marines, three companies of Washington, D.C. volunteers, and the men from the Ordnance Department; 348 men in all. Security concerns for the navy yard eased on 27 April when the 71st New York Infantry Regiment marched in and took up defensive positions. At Dahlgren's suggestion, unit officers made their quarters in the Commandant's House, while he took up residence in his office near the waterfront. There the busy officer spent most of his time in the next few years.

The destruction of the Norfolk Navy Yard, then the Navy's largest facility, and its huge reserves of naval ordnance to prevent their capture by southern troops, added to the Washington Navy Yard's importance as an arsenal and a base for operations in the Potomac River and Chesapeake Bay. Dahlgren kept his shops open around the clock; each day they produced 200 shells, 25,000 percussion caps, and 35,000 Minie and musket balls, in addition to casting, rifling, finishing, and mounting boat guns.[17]

The Union Navy's Potomac Flotilla, which operated from the navy yard, prevented Confederate forces from using the river and its tributaries, enabling northern generals to move their forces the length of the waterway. Recognizing the importance of the relationship between the Washington Navy Yard and the

Above. *"Honorable Gideon Welles, Secretary of the Navy" by J.M. Butler (Navy Art Collection). The dynamic and talented Welles led the Navy during the Civil War*

Left. *The 71st New York Infantry Regiment in formation at the navy yard during the Civil War.*

"The Contest for the Henry Hill" (Navy Art Collection, artist unknown). This action took place at the First Battle of Bull Run.

THE CONTEST FOR THE HENRY HILL.

Potomac Flotilla, the Navy assigned both commands to one officer.

One of the war's first waterborne expeditions was launched from the yard on 23 April 1861 when steamers *Baltimore* and *Mount Vernon* embarked the 1st Zouave Regiment under Colonel Elmer Ephraim Ellsworth and transported them the short distance to Confederate-held Alexandria, Virginia. The northern troops occupied the town and captured thirty-five of its southern defenders the next morning, but when the colonel removed a Confederate flag from one of its houses, the owner shot him dead. The ships carried the officer's body and prisoners back to the yard. The President and Mrs. Lincoln came down to the yard to view the fallen colonel. The Commander in Chief, who would be deeply affected by the suffering of the American people during the Civil War, graciously directed that Ellsworth's body be transferred to the White House so citizens of Washington could pay their respects.

Ellsworth's fate was shared by thousands more of his countrymen in the disastrous First Battle of Bull Run. Days after the battle on 21 July, navy yard vessels transported Union soldiers, some wounded and many demoralized, from the Virginia shore to Washington. A pitiful remnant of the 71st New York Infantry Regiment straggled back to their barracks in the yard. There was panic in the streets as Washington braced for an expected Confederate assault on the capital. Marines and sailors helped the government restore order after Bull Run by guarding, either at the yard or on board vessels, the few hundred soldiers from New York and Maine who had refused to take orders.

Dahlgren also dispatched more than 100 sailors and marines, along with three 9-inch guns and five howitzers, to newly named Fort Ellsworth in Alexandria. Eventually, 600 sailors manned the Naval Battery and defenses of Fort Ellsworth. Dahlgren also established a training program at the yard to improve the gunnery skills of the best seamen assigned to the fleet.

The Navy aided the cause in other ways. Soon after the outbreak of hostilities, slaves began fleeing southern plantations and seeking refuge on board Union warships. In August 1861, *Resolute* returned to the navy yard with seventeen African Americans. In contrast to the previous practice of returning slaves to their masters, the Navy Department told

An African-American's Reflections

On Working in the Washington Navy Yard, 1813–1865

The workers of the Washington Navy Yard—slave and free, civilian and military, American born and immigrant—have long made vital contributions to the nation and the Navy. One such worker, Michael Shiner, toiled for his country, first as a slave and then as a freed man, in the Washington Navy Yard for fifty-two years. His handwritten observations covering 1813 to 1865 appear in an oversized diary subtitled, "The Early History of Washington, D.C."

Shiner's diary opens with a compelling account of the British invasion of Maryland and attack on Washington in 1814. He describes how navy yard workers piloted vessels moving soldiers from the city to Bladensburg, Maryland, where the British trounced the Americans in battle. Shiner goes on to describe the destruction of the bridge over the Eastern Branch to retard the enemy army's advance. As the British troops approached, Shiner, along with a black man named John and an old white woman named Mrs. Reid, stood in awe watching.

> ...the ArmMy coming a Bove the tall gate in Washington[.]

> We heard the tread of British armmy feet... and as soon as We got sight of the British armmy raising that hill they look like red flames of fier all red coats and the stoks of ther guns painted With red ver Milon and iron Work.

When Shiner prepared to flee, Mrs. Reid questioned him, "What do you recon the British Wants With such a niger as you[?]"

Shortly afterwards, the British troops advanced across the capital searching buildings for supplies and then setting them on fire. Shiner recounts, however, that a storm extinguished many fires:

> i never shall for get that day one of the auuflys storms Which Raidge for a long time With out intermision it thunanrd lightent hailed and rained it takinng some ole houses up from oft ther foundation and BloWing them down in shelter and British armmy stood as if they Whrt on MarBle never went in shelter.

In the years after the War of 1812, Shiner often worked twelve hours a day on various jobs with yard mechanics, clerks, and sailors. Like his fellow workers, he labored on projects for the federal government outside the yard. On one occasion, he helped crew a barge transporting President John Quincy Adams on the Potomac and around Georgetown. Shiner's boss was so pleased with the crew's performance that he gave the men short-term passes to leave the yard. Shiner considered it an honor when his superiors allowed him, along with some yard mechanics, fellow laborers, and marines, to volunteer to move the cornerstone of a memorial to President George Washington. Shiner and other workers also helped fight local fires threatening buildings in Alexandria, Virginia, Senator Thomas Hart Benton's home in the District, and the U.S. Capitol library.

The weather could and often did affect worker productivity as well as damage portions of the yard. Shiner and his workmates had to contend with cold snowy winters, and the frozen waters of the Eastern Branch and the Potomac River. Strong northwest winds on 16 and 17 December 1831, for example, shattered six or seven 12-by-18-foot windows, leading Commodore Isaac Hull to close the yard. An eclipse on 12 February 1831, lasting from 11:30 a.m. to 3:30 p.m., halted work. Describing an especially severe drought, Shiner wrote:

> We had a smart of rain in May and din JUne they wher But little rain and in July it wher hot and dry and evry thing on the earth sufering for vegetation and no rain in august [.] [T]he ground wher so hot that the heat would exstend trough the souls of your shoes particelar thin souls and we hadent anno untel the Middele of september 1838.

He documents incidents which show that the Washington Navy Yard was not a safe place to work. Although he narrowly missed being maimed or killed on several occasions, his diary is replete with reports of injuries and deaths on the job. One German immigrant employee working on a pile-driving project was decapitated in June 1833. A shell detonated in June 1842 killing Navy gunner Thomas Barry, and the explosion of a wrought iron cannon on board U.S. steam frigate *Princeton* in February 1844 killed many government officials and a young black man. A Baltimore brick layer fell thirty feet off a scaffold to his death in 1853.

Shiner was present on numerous occasions when presidents, cabinet members, congressmen, delegates from Indian tribes, and foreign dignitaries visited the navy yard. He also witnessed many ceremonies, especially ship launchings. Because he spent so many years there, he saw the physical and technological changes in the yard, including the construction of the rolling mill plant, the ordnance shop, and the copper rolling mill engine, the first installation of gas pipes, the use of hydraulic power, and the extension of the east wall.

As a slave, Shiner had to obtain a pass from his master, the clerk of the yard, when he wished to leave the facilities, as he often did to attend religious services in ward six. The navy yard worker worshipped at a Methodist church whose pastor, Reverend Payton, attracted Shiner because "his walk and conversation proved [that] he carried the love of god in his heart." Shiner had an active social life, frequenting black-owned and -operated establishments, particularly a bar and grill on Capitol Hill. He sometimes drank too much and wound up fighting other patrons. These altercations landed him in jail where he was restrained in leg irons. Although his master seldom punished him for these transgressions, once when Shiner said he wanted to visit friends in the country and then went somewhere else, the master refused to give him a pass for a month.

The Washington Navy Yard and its workers were vital to the success of the United States Navy during the Civil War. In the dark, early days of the conflict, Shiner helped transport scarce provisions to the U.S. Capitol and prepare the yard's defenses. Like all but a few of his fellow workers, Shiner took the oath of allegiance to the United States administered by Justice William Clark at the flagpole in the navy yard on 1 June 1861. He and many other African-Americans who served the Navy helped bring victory to Union arms in the Civil War and rid the country of the scourge of slavery. Shiner does not specify when he obtained his freedom. It may have happened on 12 April 1862 when Congress abolished slavery in the District.

Shiner's handwritten account is important because it provides the insight of an African-American man who lived through a tumultuous time in U.S. history when few of his race could read or write. His keen powers of observation, sense of humor, and appreciation of events in which he took part preserve a unique perspective on the early history of the nation's capital and the Washington Navy Yard.

Regina T. Akers

This lithograph published during the Civil War shows the navy yard's two shiphouses, the experimental battery (left foreground behind a sailing vessel), ordnance manufacturing shops, and the Commandant's Office (Building 1) fronted by an ordnance display park.

Abraham Lincoln, 16th President of the United States and the nation's leader during the most trying time of its existence. Lincoln often sought relief from the pressures of his office by visiting the navy yard.

Dahlgren to release them. And, as the war made increasing demands on the North's manpower, the Navy employed black men in its service. The Secretary of the Navy authorized the enlistment of blacks, for which they received pay of $3 a month and one ration. On 16 April 1862, Congress passed the District of Columbia Emancipation Act, abolishing slavery in Washington. In the words of one historian, "freedom did not come with expanded job opportunities, but it did mean that Michael Shiner and other workers in Washington could receive a salary or, in some cases, higher pay for their labor."[18]

As the pressures of war mounted, the President sought relief by visiting Dahlgren, who soon became a confidant, and the navy yard, because Lincoln "simply liked gadgets, weapons, and munitions, and these things abounded" there.[19] On one occasion, the President picked up a breechloading carbine being tested and fired a shot. Lincoln, the same age as Dahlgren, regularly appeared unannounced at his friend's office for "coffee, cigars, and a chat with his favorite naval officer."[20] When several captains queried the Navy Department about replacing Commander Dahlgren as commandant of the navy yard, a position normally held by a captain, the Commander in Chief weighed in. Lincoln observed that Dahlgren defended the navy yard "when no one else would, and now he shall keep it as long as he pleases."[21]

African-American sailors mend their clothes on the deck of a Union warship during a break from duty.

Even as the threat to Washington from Confederate armies receded, the enemy's naval arm mounted a challenge. On 8 March 1862, the Confederate ironclad CSS *Virginia*, built on the hull of *Merrimack* and commanded by former Commandant of the Washington Navy Yard Franklin Buchanan, boldly attacked the Union blockading flotilla positioned in Hampton Roads off Newport News, Virginia. The Confederate casemate ironclad attacked sloop *Cumberland* with a 1,500-pound cast-iron ram bolted to her bow and tore apart frigate *Congress* with gunfire. Both wooden-hulled ships and many of their crewmen sank beneath the waves. The steam frigate *Minnesota* then ran aground trying to fight *Virginia*. The other vessels of the blockading force sought safety under the guns of nearby Fort Monroe.

The reaction to the disaster at Hampton Roads was nothing short of panic when word reached Washington the following day. When President Lincoln's cabinet gathered to consider a course of action, Secretary of War Edwin M. Stanton and others expressed the fear that *Virginia* would soon appear in the Potomac and begin shelling the Capitol and the White House. The entire Union cause was seen at risk from the unconventional, but single enemy warship!

The President immediately went to the navy yard to consult with Dahlgren. The President and

Dahlgren returned to the White House where they joined the secretaries of State, War, and Navy for another meeting on the crisis. Lincoln and Stanton, periodically looking through the window to see if *Virginia* had arrived at the capital, became increasingly agitated. Secretary of the Navy Welles and Dahlgren tried to assure the assemblage that Kettlebottom Shoals would probably prevent the ironclad from ascending the Potomac to Washington.

At Dahlgren's suggestion, the President authorized *Wabash* to join defensive forces near Hampton Roads. The commander also proposed posting a steamer on lookout duty south of the city. The Secretary of War did not consider these actions sufficient. Stanton insisted adamantly that sixty barges be loaded with stone and sunk in the river to block the channel. With equal vehemence, the Secretary of the Navy fought the proposal. As a compromise, the President ordered that the vessels be prepared and positioned in the river but not sunk unless the enemy warship made its appearance.

The crisis passed as quickly as it had arisen when the Union ironclad *Monitor* reached Hampton Roads from New York. After a blazing, close-order battle that lasted four hours, the battered Confederate vessel retired from the scene, never to fight again. She was later scuttled to prevent capture. Months after the battle, as a naval party with the President on board made its way down the Potomac, someone asked about the purpose of the barges lined up along the Maryland shore. Lincoln responded: "Oh, that is Stanton's Navy. That is the fleet concerning which he and Mr. Welles became so excited in my room. Welles was incensed and opposed the scheme, and it has proved that Neptune was right. Stanton's Navy is as useless as the paps of a man to a suckling child."[22]

Later in the war, when *Monitor* put in at the navy yard for repairs, the President visited the victor of

Above. *Colonel Ulric Dahlgren, the bold and courageous son of John and Mary Clement Dahlgren. When Ulric's leg had to be amputated because of wounds sustained in battle, the limb was entombed behind a memorial plaque placed on a new building at the navy yard. The leg has since disappeared.*

Left. *"The Ironclads" by Raymond Bayless. CSS* Virginia *is in the foreground, and USS* Monitor *in the background, right. Courtesy of Raymond Bayless*

This hand-tinted woodcut depicts fishermen working in the Eastern Branch, the shiphouses and shops of the Washington Navy Yard, and the unfinished U.S. Capitol building during the momentous April of 1861.

the Battle of Hampton Roads. This visit reinforced Lincoln's view that the Navy should develop a fleet of ironclad vessels able to withstand heavy shot and shell. Dahlgren exploited the President's support to press for the establishment of facilities at the yard for the testing and production of quality armor plate for warships. Congress, however, did not approve the proposal. Dahlgren also conferred a number of times at the yard with *Monitor's* designer, John Ericsson, about optimum ordnance and armor plating for Union vessels.

The sailors of the Washington Navy Yard once again answered the call when Confederate Lieutenant General Thomas "Stonewall" Jackson's forces threatened Harpers Ferry in the spring of 1862. Acting Master Daniels and cavalry Colonel Ulric Dahlgren, the commandant's son, deployed to Maryland with a company of thirty-two seamen armed with four cannon, two howitzers, and one Dahlgren 9-inch gun. When Confederate Lieutenant General Jubal Early's forces threatened Washington later in the war, 800 sailors assigned to the navy yard reinforced Fort Lincoln to help defend the city against an attack that never came. Early retired south rather than test what had become a formidable defensive redoubt.

The bold and talented Dahlgren, who ever longed for glory in battle, pressed his superiors for a combat command. They refused to part with the services of this brilliant ordnance engineer, administrator, and presidential advisor. In fact, in July 1862, when Welles divided the Navy's administrative responsibilities among a number of newly established bureaus,

Above. *Hundreds of smoothbore cannon, rifled guns, and mortars and stacks of roundshot were photographed by Mathew Brady in the gun park of the navy yard.*

Left. *A Mathew Brady photograph of the navy yard's main ordnance foundry, which stood until World War I.*

Above. Sailors man the yards in salute to President Lincoln during the visit of Brazilian steamer *Paraense* to the navy yard in January 1863.

Right. A Mathew Brady photo of ordnance stores and armor test pieces on the west side of the yard. The cupola in the background still stands.

he appointed Dahlgren as head of the Bureau of Ordnance and allowed him to continue his experimental and testing work at the yard.

Adding to Dahlgren's frustration over his failure to secure a combat command early in the war was concern for his son Ulric who had lost a leg in a skirmish on 6 July 1863, just after the Battle of Gettysburg. To honor the officer, who lay close to death from loss of blood in the Commandant's House, his severed limb was sealed in the wall of a new foundry building (208) at the navy yard and a plaque placed over the spot. Mysteriously, the leg could not be found when the bizarre crypt was opened years later.

Ultimately, John Dahlgren was promoted from commander to captain and then rear admiral and given a combat assignment, command of the naval forces besieging Charleston, South Carolina. Sadly for Dahlgren, this duty did not bring the ordnance expert the glory he desperately sought. The conflict demanded one more sacrifice of the dedicated and patriotic naval leader. Ulric was killed leading a cavalry raid near Richmond.

Dahlgren, his sailors, and the skilled and semi-skilled workmen of the Washington Navy Yard were vital to the success of the Union war effort and loyal citizens of the United States of America. At the outbreak of the war, naval leaders were concerned that many workers harbored southern sympathies. In May and June of 1862, however, 917 sailors and civilian employees swore an oath of allegiance to the United States in a public ceremony at the yard. Only eight workers refused to take the oath and were fired.

Usually without overtime compensation, skilled workers often toiled on Sundays and for two extra hours each weekday. Even though Congress passed legislation to decrease the difference in pay between workers at government navy yards and those at commercial shipyards, this proved hard to implement. Navy yard workers paid a price in lower wages and longer hours for their dedication to government service. Moreover, some politicians expected loyalty of another sort from navy yard employees. Throughout the war and with varying success, influential congressmen pressed the Navy Department to hire their political supporters.

John Wilkes Booth, the assassin of President Lincoln.

As often happens in wartime, the enormous outpouring of funds from government coffers for provisioning new ships and purchasing weapons, supplies, and other necessities of war stimulated corruption. In 1863, Secretary Welles relieved from duty Commodore Andrew A. Harwood, Dahlgren's successor as commandant, for his inattention to the misuse of government funds by some employees. Welles also fired the naval storekeeper and two master workmen for their illicit involvement in the sale of scrap copper.

When news of Lee's surrender at Appomattox reached the people of Washington on 10 April 1865, navy yard workmen fired salutes with Dahlgren guns and joined other Washingtonians and military bands who marched joyously through the streets of the national capital. Navy yard carpenters and joiners helped decorate the Navy Department building for end-of-war celebrations. All over the capital, flags, bunting, and lighted candles were displayed to mark the happy occasion. Sadness replaced joy on 14 April, when actor John Wilkes Booth assassinated President Lincoln, a frequent and welcome visitor to the navy yard, who was beloved by many of the sailors and workers there.

Booth escaped into Maryland via the navy yard bridge (now the 11th Street Bridge). Between 17 and 29 April, the eight men suspected of conspiring with Booth to kill the President were brought to the navy yard. They were put in irons on board monitor *Saugus* and other warships in the river and kept under constant guard. Secretary of War Stanton directed that navy yard officials put a canvas bag, with only a hole for eating and breathing, over the head of each man. The guards also kept a suicide watch over one of the accused men.

Early on the morning of 27 April, federal authorities brought Booth's body to the yard and placed it on board monitor *Montauk*. That day, the Army's Surgeon General carried out an autopsy on Booth in

In April 1865, Army Department officials autopsied the body of a man they claimed was John Wilkes Booth on board monitor Montauk, *the vessel on the left in the photo and similar to monitor* Miantonomoh *in the foreground.*

Raphael Semmes, the daring Confederate naval officer confined at the navy yard in 1865 and 1866.

the presence of an assistant, various Army officers, and a photographer. Even though these activities occurred at the navy yard, under Navy Department jurisdiction, Stanton took personal interest in the disposition of the assassin's remains. Without the Navy's knowledge or permission, Army Department representatives removed the body from the yard that afternoon. This action and Stanton's earlier involvement with Booth's apprehension and death in Virginia have fueled speculation to this day that the Secretary of War was somehow connected with the conspiracy. Some historians even doubt that the remains held at the navy yard were those of Booth.[23]

The next act in this assassination drama took place a few days later, when an Army guard transferred the eight prisoners from the navy yard to the federal penitentiary at what is now Fort Leslie J. McNair. Three of the men and another conspirator, Mary Surratt, were hanged there shortly afterward. Finally, on 4 May 1865, the guns at the navy yard fired a salute every half-hour from noon to sunset to mark the burial in Springfield, Illinois, of President Abraham Lincoln.

The Post-Civil War Era

Anger over Lincoln's assassination and the bitterly fought Civil War influenced not only Congressmen but Navy Department leaders during the Reconstruction period. In December 1865, Secretary Welles directed the arrest and confinement at the navy yard of Rear Admiral Raphael Semmes, the commanding officer of Confederate raiding ships *Sumter* and *Alabama*, which had cruised the oceans of the world sinking many Union merchant vessels. The naval officer was charged with various offenses, but the real reason, according to Navy Judge Advocate General John A. Bolles, was that

Lieutenant William N. Jeffers on board Monitor *during the Civil War. As Chief of the Bureau of Ordnance during the 1870s, Jeffers had a major impact on the development of rifled naval ordnance.*

"the exploits of Semmes were by no means forgotten by any one; and they rankled in the memories of thousands who had suffered from the depredations of the *Sumter* and the *Alabama*."[24] Northern newspapers whipped up the public by reporting that Semmes had harshly treated prisoners and inspired other misdeeds. A long prison term or even execution of the former Confederate naval officer were possible outcomes of a trial in 1865.

During the four months it took the Navy to investigate the allegations against Semmes, he spent his hours confined under marine guard in the attic of a navy yard building. His one pleasure was viewing the Maryland and Virginia shores from his window. Semmes wrote in a diary that his captors fed him satisfactorily and provided him daily with newspapers and "plenty of water, soap, and towels."[25] In April 1866, when Bolles finally decided that the charges were groundless and the public's rage had subsided, he persuaded Welles to release Semmes, who returned to Alabama and retirement.

One of the last acts of the Civil War drama occurred in February 1867, when the steam sloop-of-war *Swatara* took on board in Egypt a Lincoln assassination conspirator, John Harrison Surratt, who had fled to first England, then Italy, and finally Egypt. When apprehended, Surratt was found dressed in the uniform of a Zouave of the Pope's army. *Swatara*'s captain, Commander William N. Jeffers, clapped Surratt in irons and confined him to the brig for the transatlantic voyage. The day after *Swatara* anchored in the Anacostia River, Jeffers had the prisoner rowed to a navy yard wharf and turned over to a U.S. marshal accompanied by District police.

The postwar years were hard on the Navy. A fleet of 700 commissioned vessels shrank to a force of just 52 vessels between 1864 and 1870.[26] The once mighty Potomac Flotilla almost disappeared as its monitors, gunboats, and other vessels were sold, scrapped, or left to rot in the Anacostia.

With the war emergency over, the Navy Department tried to reduce the number of workers on the navy yard payroll. Congressmen, especially the legislators from Maryland, demanded special consideration for workers from their districts. Secretary Welles successfully resisted some but by no means all efforts to influence the hiring and firing of navy yard employees, and he had to pay constant attention to the matter. To bring some uniformity to the employment process, he established a preference system for the Navy and Army veterans who made up about one third of the yard's workforce by the fall of 1865. Political influence, favoritism, and corruption—especially after Welles left government and the administration of President Ulysses S. Grant came in—ran rampant at the Washington Navy Yard, as it did in the nation's other navy yards. Party bosses, working through navy yard managers and master craftsmen, demanded and received financial contributions and election support from workers as payment for their jobs. Navy regulations and the efforts of successive commandants did little to curb these abuses.

Unfortunately, the Navy Department was also guilty of questionable actions with regard to the political process. In 1869, when Radical Republicans ruled in Congress, Vice Admiral David Dixon Porter, a Civil War hero, issued a directive with the blessing of the Secretary of the Navy that called for all navy yard commandants to ensure that their workmen were loyal to the Grant administration. He stipulated that in employment "preference in all cases should be given to those who have belonged to the Union party."[27]

The post-Civil War years witnessed the material decline of the Washington Navy Yard and the Navy itself. Americans wished to recover from the conflict's enormous expense in blood and national treasure. With no threat to the nation likely at home or from abroad, Americans focused on building a prosperous industrial society and exploiting western agricultural lands and mineral resources.

As a result, Congress drastically reduced appropriations for the Navy. The Navy did not build a single

John Surratt, accused plotter in the assassination of Lincoln, on his arrival at the navy yard in 1866 (U.S. Naval Institute).

armored vessel for twenty years after the war. Reduced labor forces continued to work with sail, steam, and wood rather than with newer naval technologies, many developed in the United States and adopted by European powers for their naval forces. The Civil War monitors, gunboats, and wooden-hulled sailing ships got older and increasingly decrepit.

John A. Dahlgren, who once again served as Commandant of the Washington Navy Yard, from 1869 to 1870, was a metaphor for this melancholy period of naval history. He had lost his health and his son Ulric to the war. And the brilliance of his reputation as a weapons designer had faded as newer designs gained favor. Dahlgren, whose name was synonymous with the Washington Navy Yard for many years in the 19th century, died from a sudden heart attack in the Commandant's House on 12 July 1870.

Much of Dahlgren's pioneering work with ordnance was carried on by Captain William Jeffers, who was assigned duty as the assistant to the Chief of the Bureau of Ordnance at the Washington Navy Yard on the death of his illustrious predecessor. Jeffers, editor of the well-regarded *Ordnance Instruction for the U.S. Navy*, pushed for the manufacture by the Navy of rifled guns to replace the smoothbore weapons favored by Dahlgren. Because of the latter's opposition and lack of funding support for new ordnance, Jeffers' recommendations were not heeded during his eight-year tenure with the bureau. But Jeffers' "strong and effective leadership [provided] the principal impetus for most of the improvements in material and expertise in the seventies" that paid benefits in the 1880s when the Navy began rebuilding.[28]

He did oversee the rifling of the old smoothbores in the Navy's arsenal that gave the guns greater accuracy and range. At his urging, the Navy adopted breech-loaded boat guns and the development of officers skilled in ordnance design and production. Jeffers also helped inspire American steel manufacturers to develop the equipment and processes for producing the large, breech-loaded guns that would arm America's steel navy of the late 19th century. By the time he retired, his ideas about ordnance had taken hold. In the words of one scholar, "Jeffers' foresight and persistence during a so-called period of decay and stagnation resulted in substantial improvement in the quality of domestic gun steel, in knowledge of information about gun design, and in professional expertise that contributed significantly to the resurgence of American naval power during the eighteen eighties." He added that, "with the exception of Dahlgren, Jeffers had "more impact on the development of naval ordnance than any other who held the position of bureau chief."[29] Much of Jeffers' work was done at the Washington Navy Yard.

Gunboat Nipsic *(or possibly* Yantic*), the last naval vessel built at the Washington Navy Yard, tied up in front of the west shiphouse in the early 1870s.*

USS Atlanta, *one of the new steel ships that entered the fleet in the late 19th century.*

Arming a New Navy

The Navy's fortunes improved in the 1880s as the nation once again paid attention to international affairs and considered rebuilding its military strength. Foreign naval forces increasingly appeared in the Caribbean and the Pacific as European governments sought overseas colonies and global influence. Americans too began to consider outlets abroad for their manufactured goods, farm produce, and missionary zeal. American nationalists concluded that since the United States was fast becoming a strong nation, the Navy should be technologically advanced and militarily powerful. In 1881 and 1882, Secretary of the Navy William H. Hunt and his successor, William E. Chandler, recommended to Congress the construction of four, all-steel warships. Within a few years of their authorization, cruisers *Atlanta*, *Boston*, and *Chicago* and despatch vessel *Dolphin*—the ABCD ships—put to sea, inaugurating a new era in naval history.

In 1883, President Chester A. Arthur asked the Secretary of War and the Secretary of the Navy to appoint a Gun Foundry Board and have it investigate the optimum approach to ordnance design and manufacture. The board members spent the next year visiting steel mills in the United States and gun foundries in Great Britain, France, and Russia. On their return, they recommended establishment of separate gun foundries in the United States for the Army and the Navy.

Based on the board's investigation, in April 1886 Secretary of the Navy William C. Whitney directed that all Navy ordnance manufacturing be carried out by the Washington Navy Yard. The Naval Gun Factory, formally established on 14 August 1886, slowly took shape as the Navy enlarged existing buildings and fitted them with heavy-lift cranes, laid tracks to connect the yard with the Baltimore and Ohio Railroad, and installed telephone lines. To strengthen the gun factory's relationship to the fleet, officers and sailors were trained at the yard in ordnance design, production, operation, and repair. By 1892, the almost 1,000 yard workers (since 1891 hired on a merit system) were manufacturing 4-inch, 6-inch, 8-inch, 10-inch, and 13-inch guns and the shells they fired.

In addition to heavy guns for its ships, the new Navy needed steel plate armor for their protection. Hence, the gun factory's Lieutenant William Jaques worked with British industrialists and John Fritz of the Bethlehem Iron Company to establish at the factory in Bethlehem, Pennsylvania, the United States' first armor plate forging plant.

As they had in other areas of naval technology during the latter half of the 19th century, European powers surged ahead of the United States in the design and testing of warship hulls. The British, Italians, and Russians constructed huge water-filled

Left. *Naval Gun Factory workers add finishing touches to the guns that will be used by naval landing parties.*

Below. *Under their instructor's watchful eye, sailors in the Seaman Gunners Class work the components of a large gun on the navy yard waterfront.*

Right. *Workers of the pattern and joiner shop pose in front of their workplace for this 1905 photograph. From the earliest days of the navy yard, many residents of Washington, D.C., earned their living at the naval facility.*

Below. *Construction began in February 1897 on the Experimental Model Basin, where the Navy would test the hull designs of new steel battleships, cruisers, and other warships entering the fleet. To the left of the construction is Building 1, and to the right, the east shiphouse.*

basins to test hydrodynamic properties of less-than-full-size models of proposed hulls. By towing such models through these basins, marine engineers could scientifically determine the top speed of a proposed full-size warship. Determined to develop a fleet of fast, modern steel warships, in June 1896 Congress passed and President Grover Cleveland signed a bill authorizing construction of a hull-testing basin at the navy yard.

Work on the Experimental Model Basin (EMB), as the facility became known, was begun in February 1897 under the direction of Naval Constructor David W. Taylor. Navy pile drivers embedded pilings deep into the soil near the navy yard waterfront. The Penn Bridge Company of Beaver Falls, Pennsylvania, removed soil and reinforced the pilings with thick layers of concrete, broken stone, and asphalt. The contractor then built a basin that was 470 feet long, 42 feet wide, and almost 15 feet deep and could hold a million gallons of fresh water provided by the District water system. The constructors covered the basin with a 500-foot-long, brick building crowned by a long skylight. A 40-ton towing carriage built by the

Above. *Rear Admiral David W. Taylor, a pioneer of hydrodynamics in the United States.*

Left. *Managers and workers watch as water fills the newly built Experimental Model Basin in the early summer of 1898. For the next half-century, the Navy tested thousands of hull models at the state-of-the-art facility.*

Right. *A civilian scientist prepares a hull model for testing in the water-filled model basin.*

Below. *The Experimental Model Basin (Building 70) just before its dedication on 4 July 1899. The shiphouse to the left, a waterfront landmark for much of the 19th century, was demolished in 1901.*

"USS Maine Blowing Up in Havana Harbor on 15 February 1898" (Navy Art Collection, artist unknown).

William Sellers Company of Washington rested on two pairs of steel rails running parallel to the basin. A reciprocating steam engine that drove three General Electric dynamos and four motors powered the carriage, allowing it to tow a model at speeds ranging from 0.5 to 18 knots. When Taylor announced completion of the facility on Independence Day of 1899, the Experimental Model Basin was the largest such facility in the world.

The nation went to war with Spain as the model basin was under construction. But the Spanish-American War of

Rear Admiral William T. Sampson.

1898 ended so quickly that the current operations of the Naval Gun Factory had little effect on the outcome of the conflict. The ordnance facility proved its worth, however, when the ships of the U.S. Navy, armed with guns manufactured mostly in Washington during the 1880s and 1890s, devastated Spanish flotillas in the battles of Manila Bay and Santiago de Cuba. Guns produced during the war by the 2,255 workers, who toiled in shifts in a 24-hour, 6-day work cycle, entered the fleet in great number in the years afterward.

There was another connection between the navy yard and the Spanish-American War. Rear Admiral William T. Sampson, commander of the North Atlantic Squadron, and Commodore Winfield Scott Schley, commander of the Flying Squadron and Sampson's subordinate, engaged in a nasty postwar debate in the press over each other's wartime performance. In 1901, at Schley's request, Secretary of the Navy John D. Long convened a court of inquiry in the Sail Loft over the Gunner's Workshop at the Washington Navy Yard. The press and the public, including yard workers during lunch breaks, thronged the proceedings, which lasted from September to December. In the end, Admiral George Dewey, the hero of Manila Bay and President of the Court, spoke for the other officers when he concluded that Schley had issued misleading reports and had not followed orders promptly or with great professional skill. President Theodore Roosevelt rejected Schley's appeal to reverse the court's judgment. Above all, the President wanted to end the public squabble between naval leaders, for he had big plans for the U.S. Navy in the new century.

Convinced by Roosevelt, the far-thinking strategist Alfred Thayer Mahan, and other "navalists" that the new international stature of the United States demanded a powerful battle fleet, Congress appropriated huge sums of money to build one. From 1901 to 1905, the nation supported the construction of ten battleships, four armored cruisers, and seventeen other warships. The Naval Gun Factory once again worked on a 24-hour schedule, employing more than 2,000 workers to produce the hundreds of guns that soon joined the fleet. Directing the factory during this enormous growth period was Captain E.H.C. Leutze.

For the first time in the Navy's history, the yard began to rely on other government and commercial plants to produce much of the ordnance needed for the new fleet. The Navy Department enlisted the services of outside contractors and the Army's Watervliet Arsenal, which manufactured guns to Navy specifications, to keep up with the building schedules.

Sailors at the navy yard prepare a casket for a March 1912 ceremonial procession through the streets of Washington to Arlington National Cemetery. The fallen sailor and many of his shipmates died in the explosion of battleship Maine.

The practice of using private industry to produce naval guns increased to the point that in 1905 the Washington facility returned to an 8-hour work day and reduced its workforce to 339 men. The Naval Gun Factory, however, continued to be the Navy's primary site for ordnance design and testing and a center of technical excellence. This factor became increasingly important as Commander William S. Sims and other innovative naval officers pressed for the fleet's adopting improved gun sights and fire-control equipment for long-range shooting. Accurate gunfire was critical to the combat success of 20th-century "Dreadnought" battleships.

President Theodore Roosevelt, his successors, and U.S. naval leaders of the new, powerful fleet desired to put to sea in their own "flagships." During the late 19th century, former steamer *Despatch* and the steel-hulled *Dolphin*, operating from the Washington Navy Yard, had occasionally transported presidents and secretaries of the Navy on diplomatic voyages and Potomac River excursions. Beginning in 1898, however, the Navy formally assigned vessels to the Washington Navy Yard to serve the diplomatic and social needs of the President, the Secretary of the Navy, and other high-ranking government officials. Between 1902 and 1929, converted yachts *Sylph* and *Mayflower* alternated in that duty. Roosevelt and President William Howard Taft boarded the vessels at the yard for day-long outings in the vicinity of the nation's capital, visits to the fleet in the Atlantic, or cruises to New York and New England. The elegantly furnished *Mayflower*, a two-masted vessel displacing 2,690 tons and capable of making 17 knots, served as the site at Portsmouth, New Hampshire, for diplomatic negotiations ending the Russo-Japanese War.

President Roosevelt's decision to send what came to be called "The Great White Fleet" around the world to highlight America's new prominence on the international stage compelled the Naval Gun Factory to rush the completion of needed ordnance equipment. In the three months before the Atlantic Fleet departed from Hampton Roads, Virginia, in 1907 the gun factory provided the warships there with 146,000 pieces of manufactured ordnance, including more than 11,000 spare parts, 432 telescopes, 314 gun accessories, 264 gun carriages, 172 gun sights,

156 sets of training gear, 150 firing locks, and 130 guns. The gun factory shipped additional material to San Francisco while the fleet, under Rear Admiral Robley D. "Fighting Bob" Evans, made its passage through the South Atlantic, around Cape Horn, and north along the Latin American coast to the U.S. West Coast. When the fleet steamed from San Francisco en route to New Zealand, Australia, the Philippines, China, and Japan in 1908, it was well armed for the historic, around-the-world cruise.

Significant improvements were made to the gun factory during the years before World War I. Oil gradually replaced coal as the fossil fuel to fire boilers. The Bureau of Ordnance installed an 80-ton crane in the gun shop and enlarged the building to handle the huge 14-inch naval rifles the facility began to produce. Electric and gasoline-powered vehicles replaced the teams of oxen that had labored for many years to move guns and machinery around the yard. The Navy also put into operation a new foundry, at the time one of the most advanced U.S. facilities, which soon manufactured castings for steel, iron, and bronze ordnance.

The Washington Navy Yard assisted in the birth of U.S. naval aviation. On 14 November 1910, civilian contract pilot Eugene Ely flew his Curtiss biplane from a wooden deck installed on cruiser *Birmingham*, anchored in Hampton Roads. The following month, the Chief of the Bureau of Ordnance directed the navy yard to design and build a catapult for launching airplanes from the deck of a warship. Several attempts to launch aircraft with the device failed, but on 12 November 1912, a modified catapult propelled Lieutenant T. G. "Spuds" Ellyson and his plane into the air.

Like many boys before and since, these visitors pose for the camera on the barrel of a gun displayed at the navy yard.

The Early Days of Naval Aviation

The Washington Navy Yard became closely associated with the development of naval aviation soon after farsighted strategists and aircraft designers concluded that a lighter-than-air craft, an airplane, could revolutionize warfare. Within a few years of Orville and Wilbur Wright's historic flight at Kitty Hawk, North Carolina, on 17 December 1903, innovative naval officers reasoned that aircraft could help the fleet's battleships deliver accurate fire at great distances by spotting where rounds hit the water. They also understood that planes launched from ships could locate and then bomb an enemy fleet far over the horizon. Because the Washington Navy Yard was one of the service's principal design, testing, and production plants, and was close to powerful political leaders in the nation's capital, aviation enthusiasts selected the facility to investigate promising concepts and develop new equipment.

Instrumental to this effort was Lieutenant Holden C. Richardson, the Navy's first aviation engineering and maintenance officer, who was assigned to the navy yard on 10 October 1911. He worked for another aviation pioneer, Captain Washington I. Chambers, who was bursting with ideas. In a letter dated 16 October 1911, Chambers described a promising test of the new Forlanini hydrovanes for use as floats with seaplanes. The very next day he sang the praises of heavy oil (or diesel) engines and turbine engines to Glenn H. Curtiss, the soon-to-be-famous designer and manufacturer of airplanes for the Navy, observing that "this turbine is the surest step of all, and the aeroplane manufacturer who gets in with it first is going to do wonders."

On 31 July 1912, off Annapolis, Maryland, Lieutenant Theodore G. "Spuds" Ellyson piloted a Navy Curtiss A–1 as it was catapulted from the deck of a warship in the first attempt at such a launch in naval aviation history. The Naval Gun Factory had built the catapult, a device powered by compressed air, based on an initial design drawn up in Chambers' office in the Navy Department. Midway through this first launch, however, a gust of wind caught the plane, which was not secured to the catapult, and pitched it into the water. Despite its initial failure, the Navy refined and standardized the methods of catapult launching, forever changing the nature of naval warfare.

President William Howard Taft was so impressed with the work of Chambers, Richardson, and the Naval Gun Factory that he readily endorsed the recommendation of the Secretary of the Navy to look into establishing an aerodynamical laboratory at the navy yard. In further recognition of Chambers' unique contributions, late in 1912 the Aeronautical Society presented the naval officer with a medal for his "unusual achievements in being the first to demonstrate the usefulness of the aeroplane in navies, in developing a practical catapult for the launching of aeroplanes from ships, in assisting in the practical solution of the hydroaeroplane by the production in association with others of the flying boat, in having been instrumental in the introduction into our halls of Congress of bills for a National Aerodynamic Laboratory, and a Competitive Test, and through his perseverance and able efforts in advancing the progress of Aeronautics in many other channels."

On 9 October 1913, Acting Secretary of the Navy Franklin D. Roosevelt designated Chambers the Senior Member of a newly established Board on Naval Aeronautical Service. The board, whose other members included Richardson, Commander Carlo B. Brittain, Commander Samuel S. Robison, Lieutenant Manley H. Simons, Lieutenant A. A. Cunningham, and Lieutenant John H. Towers, was charged with planning the organization of a Naval Aeronautic Service. These men fought an uphill, but ultimately successful struggle to convince skeptical Navy leaders that aviation could play a prominent role in future operations for control of the sea. Chambers and Richardson, for instance, energetically overcame funding shortages and pushed equipment testing to advance the cause.

These were exciting years for American naval aviation. On 9 March 1914, the Navy established its first wind tunnel at the navy yard's Experimental Model Basin. Within months, the facility began testing less-than-full-size airplane

A seaplane is launched by catapult from a navy yard barge operating in the Anacostia River during tests conducted in 1926.

models at various wind speeds and directions.

Lieutenant Richardson, who earned his wings on 12 April 1915 to become Naval Aviator #13, and his shop at the navy yard continued to work on catapult designs. On 16 April 1915, a catapult fabricated by the Naval Gun Factory propelled Lieutenant Patrick N. L. Bellinger and his AB-2 flying boat into the air from a barge positioned off Pensacola, Florida; this was the Navy's first successful catapult launch of an aircraft from a waterborne vessel.

On 10 July 1915, the Secretary of the Navy authorized establishment at the yard of the Aeronautical Engine Laboratory that soon began investigating various propulsion machinery for the fighters, light bombers, and scout planes that the Navy would need when the United States went to war against Germany in April 1917. To handle the increased workload brought on by war, on 19 October 1917 the Navy established the Anacostia Naval Air Station across the river from the navy yard. By the summer of 1918, most aircraft testing took place at the air station.

After the war, aviation work continued at the navy yard, although on a smaller scale. On 13 June 1923, President Warren G. Harding toured Langley (CV 1), the Navy's first aircraft carrier, then at the navy yard. The ship became the primary vessel for testing the Navy's naval air war concepts and tactics. During the 1930s and early 1940s, the Naval Gun Factory and the Experimental Model Basin developed, tested, and manufactured a range of engines, airframes, bombs and rockets, and other aviation equipment. The Navy's requirements for aviation resources were so enormous by then, however, that other naval facilities around the country eclipsed the navy yard in terms of importance. Still, one can credit the Washington Navy Yard with a major role in the birth and early development of U.S. naval aviation.

Mark L. Evans

Above. *President and Mrs. William Howard Taft, U.S. and Argentine civilian officials, and naval officers of both nations pose for a photo on board Sarmiento when the Argentine Navy warship arrived at the Washington Navy Yard in October 1910 for an official visit to the United States. Photo courtesy of Captain Hugo Diettrich, Argentine Navy (Ret.)*

Left. *USS Mayflower (PY 1) steams on the Anacostia River past the Army War College.*

Right. *The Great White Fleet gets underway from Hampton Roads, Virginia, in December 1907.*

Below. *President Theodore Roosevelt speaks to sailors of the Great White Fleet at the end of their globe-circling cruise in February 1909.*

Left. *The early days of naval aviation. A Curtiss Pusher plane takes off from the deck of* Birmingham *(CS 2) on 14 November 1910.*

Below. *A navy yard crew prepares to test a catapult for launching the Curtiss seaplane.*

Ordnance for a Navy Second to None

The outbreak of war in Europe during the summer of 1914 did not stimulate increased production at the Naval Gun Factory. President Woodrow Wilson, the Congress, and many Americans hoped to avoid involvement in the conflict, so they did not support huge increases in military department budgets for 1914 and 1915. The gun factory continued work appropriated for years earlier, completing, for example, the triple 14-inch gun mounts for new battleships *Oklahoma, Nevada, Pennsylvania,* and *Arizona.*

This honeymoon did not last long. German submarine attacks on British passenger and merchant ships in the North Atlantic, which resulted in many American deaths, inflamed public passions. Citizens believed that if the United States built a much stronger Navy the country could deter German attacks at sea and still keep out of the war. The Naval Act of 1916 provided nearly $500 million for constructing a "Navy Second to None." The three-year program anticipated the building of ten battleships, six battle cruisers, ten scout cruisers, fifty destroyers, nine fleet submarines, fifty-eight coastal submarines, and sixteen support vessels.

The Naval Gun Factory was charged with the production of 14-inch guns for the battle cruisers as well as with the full production of the newly designed 16-inch, 45-caliber and 50-caliber naval rifles intended for the latest battleships. The hugh 16-inch gun proved to be a weapon of unparalleled range, destructive power, and accuracy. With the exception

Forging the weapons of war. Naval Gun Factory workers process steel for one of the thousands of guns needed by the U.S. fleet in World War I.

A tractor mount and 7-inch naval rifle manufactured by the gun factory are parked in front of Tingey House prior to their shipment to the battlefield in France during World War I.

of the Imperial Japanese Navy's 18-inch guns, American 16-inch rifles were the largest and most powerful warship guns ever produced and served the U.S. Navy in every war it fought during the late 20th century.

The Naval Act of 1916 demanded a multifold increase in the gun factory's industrial capacity. In the next two years, the Navy bought land on the east and west sides of the yard, filled in land on the waterfront, and hired thousands of new workers. By April 1917, the 6,000 gun factory employees were working at fever pitch in three shifts to prepare for the imminent U.S. entry into the war. By the end of 1918, the labor force had swelled to more than 10,000 workers.

Like their compatriots around the nation, these workers enthusiastically supported the war effort. The master mechanics of the shops, chief clerks of offices, and family members of navy yard employees bought $225,000 worth of War Savings Certificates and Thrift Stamps during the war. Employees held dance benefits, put their spare pennies in empty shell casings placed around the yard, and in other ways contributed an average of $3,800 each month to the Navy Department Auxiliary of the American Red Cross. On one occasion, Secretary of the Navy Josephus Daniels accepted a donation of $5,000 and made appropriate remarks at an employee gathering on the parade ground. Women of the Naval Reserve and wives and daughters of navy yard workers knitted socks and sweaters for the "gallant boys aboard our battleships and destroyers, whose constant vigil guarded our boys on their way 'over there' from the treacherous and cowardly German submarine, that never fairly fought, but always 'struck below the belt'."[30]

During the war the Navy constructed new buildings and bought advanced tools and machinery. The yard installed new electric furnaces and rebuilt and enlarged the existing open-hearth furnaces. The Experimental Model Basin installed a new wavemaking machine.

This increased capacity resulted in the wartime production of hundreds of guns, and thousands of breech mechanisms, spare parts, gun sights, and 12,000 pieces of wireless, or radio equipment, and tens of thousands of cartridge cases. Most of the mines used in the North Sea Mine Barrage, a huge minefield laid between Norway and Scotland to hinder German submarine operations, were manufactured by the gun factory, as were depth charges and depth charge throwing and dropping devices. The navy yard continued to issue torpedoes, which the facility had designed, tested, and manufactured since the 1880s for Navy surface ships and submarines. To handle the increased need during World War I, the Navy built a torpedo plant in nearby Alexandria, Virginia, and put the navy yard in charge of its direction. The output of ordnance material at the yard tripled from 1916 to 1918.

One of the gun factory's most stellar accomplishments was design and production of railway carriages on which were mounted 14-inch naval rifles. In 1917, Rear Admiral Ralph Earle, Chief of the Bureau of Ordnance, persuaded the Navy Department that these rail-mounted guns could effectively counter German guns shelling the French capital of Paris and other Allied sites from great distances. Employing workers experienced in battleship turret and rail car design and construction, the gun factory quickly developed a suitable design and contracted

with Baldwin Locomotive Works and Standard Steel Car Company of America to manufacture eight 14-inch railway mounts plus 50 locomotives and kitchen, berthing, repair, and headquarters cars.

More than 20,000 sailors volunteered to serve in Navy railway batteries, even though only a small number of men with gunnery and railroad experience could be accommodated. The Navy thoroughly trained the men at naval proving grounds in

Above. *Patriotic feeling ran high at the gun factory in World War I. Employees routinely bought "Liberty Bonds" to support the war effort when they received their wages at the installation's pay wagon.*

Right. *The Navy Yard Chapter of the American Red Cross in World War I.*

Left. *During World War I, Secretary of the Navy Josephus Daniels made inspirational remarks to navy yard workers and sailors assembled in Leutze Park and accepted their $5,000 check for war bonds.*

Below. *American sailors prepare to emplace a mine in the North Sea Mine Barrage.*

Indian Head, Maryland; Sandy Hook, New Jersey; and the Naval Gun Factory in Washington; it tested these weapons at Sandy Hook. Between April and July 1918, the first railway gun mount was completed, transported to France, tested once more, and deployed to a position opposite the German gun harassing Paris. When German aircraft spotted the weapon being moved to the front, the enemy ceased shelling the French capital and moved their own long-range weapon out of harm's way. By the end of the war, five U.S. naval railway batteries were bombarding enemy rail facilities near Laon and the road and rail links to the German forces between Verdun and Metz.

The World War I era proved to be a productive period for the Experimental Model Basin and an intellectually creative time for its guiding force, Rear Admiral Taylor. His book, *The Speed and Power of Ships*, which was based on the systematic investigations and analyses done at EMB, brought the admiral

renown in engineering circles. Its conclusions became "the basis for the Taylor Standard Series, which [guided] ship design for the next fifty years."[31] In 1914, the Navy promoted Taylor from captain to rear admiral and from head of EMB to Chief Constructor and Chief of the Bureau of Construction and Repair. Between 1914 and 1923, the admiral oversaw the design and construction of 893 vessels of all classes for the Navy.

The Experimental Model Basin tested more than 2,000 models, producing a basic body of knowledge and "attracting national and international recognition for fundamental advances in the science of hydrodynamics."[32] EMB designed and built many of the Navy's early aircraft catapults and seaplanes and established the nation's first wind tunnel to test the aerodynamic properties of airplane models.

After Taylor's retirement from the Navy, a group of the most prestigious American engineering societies awarded him the John Fritz Medal; Thomas Edison, Orville Wright, and Britain's Lord Kelvin were previous recipients. The award recognized Taylor's "outstanding achievement in marine architecture, for revolutionary results of persistent research in hull design, for improvement in many types of warships and for distinguished service as Chief Constructor of the United States Navy during the Great War."[33]

Taylor and the Experimental Model Basin made a major impact on the hull design of the Navy's 20th-century warships. One historian compared the "pre-

U.S. Navy submarine chasers search for German U-boats in European waters.

Taylor" cruiser *Olympia* "with its forward-cutting sharp bow" to the "after Taylor" *Iowa*-class battleship *New Jersey*, "with its swept-back, curved bow and a forward-jutting bulb under the water line." He observed that the "characteristic bows of later high-speed ships reflect the Taylor lines so widely that, without too much exaggeration, one could regard modern naval fleets themselves as floating monuments to his achievement."[34] Taylor rightly has been called the "founding father of hydrodynamics in the United States."[35]

Above. *A 14-inch naval rifle mounted on a specially built railway car and crewed by American sailors fires a shell toward German positions in France during World War I. The navy yard designed and assembled eight naval railway batteries.*

Left. *The model of a seaplane in the Experimental Model Basin's wind tunnel.*

Between the World Wars

Ordnance production begun during World War I continued to some degree at the Naval Gun Factory between the world wars, but in the early 1920s Congress drastically reduced naval funding. In an era of strong antiwar and isolationist feeling, some Americans hoped to limit expenditures for military forces and overseas ventures. In addition, as a result of the Washington Naval Treaty of 1922, the United States, Great Britain, Italy, France, and Japan agreed to limit the size of their fleets by scrapping ships and ending much new construction. Naturally, there was little

This contemporary drawing reflects the industrial activity continued at the gun factory in the immediate post–World War I years.

need for new ordnance. In February 1922, the navy yard discharged 1,500 workers and by the end of 1923, employed fewer than 3,000 people.

In this lean period, the navy yard concentrated on improving airplane and submarine equipment for the fleet. The Experimental Model Basin carried out aerodynamic testing of aircraft, for which the Navy showed increasing interest during the interwar years. A small staff working at EMB but under the direction of the new Bureau of Aeronautics investigated improvements to airfoil, fuselage, cockpit, and windshield designs for fighters, seaplanes, blimps, rigid airships, and kite balloons.

Following the accidental loss of several submarines during the 1920s, the Navy established an EMB experimental diving unit that worked to design better diving and salvage gear and improve underwater cutting and welding techniques. In 1927, the Navy also established the Deep Diving School in a navy yard building on the Anacostia. The sailors were taught how to operate with compressed air and an oxygen-helium mixture at depths as great as 320 feet.

As it became clear in the 1930s that the Navy would be designing and constructing a new, more modern fleet, naval leaders decided that the hull and airframe-testing facilities at the Washington Navy Yard were no longer adequate. Not only was the model basin too small but physically unsuitable. In the words of one historian, "by the mid-1930s, the basin was forty years old and, in some respects, quite decrepit. Periodic river flooding [like the flood of 1936], fire damage to the building, settlement of the foundation into the soft soil on the bank of the Anacostia, and recurrent repairs and cleaning of the aging equipment all took a toll of workdays."[36] In addition, "the basin had become technically inadequate to meet the demands of both commercial users and the rapidly modernizing Navy."[37]

So, under the driving force of Rear Admiral Emory S. Land, Chief of the Bureau of Construction and

Exceptionally Meritorious Service

Employees who have served forty (40) years or more at the Washington Navy Yard

GEORGE F. WATERS
55 years
Machinist A. R.

Has been continuously employed in the yard for 55 years. Commended for long and faithful service. Original entry June, 1865.

SAMUEL I. MILLER
40 years
Machinist A. R.

Born June 20, 1859, in the city of Washington. Entered the yard 40 years ago as apprentice, under instruction, and has served continuously during this period. Has had excellent health. Commended for faithful service during employment.

M. C. THOMPSON
40 years
Ordnanceman

Born Aug. 22, 1858, in the city of Washington. Entered the yard July 7, 1880, as Laboratory apprentice at $1.25 per day. Has been commended for his long and faithful service. Has enjoyed good health during these years.

GEOFGE S. STEWART
51 years
Leadingman

Born Sept. 1, 1851, in the city of Washington. Entered the employ of the Gun Factory on Nov. 10, 1869, as apprentice, under instructions, at $.52 per day, and has advanced by successive grades to leadingman machinist. Commended for his long and faithful service. His health has been good during this time.

JOS. M. PADGETT
77 years old
40 years

First worked in yard in 1855, Construction Department, under Samuel M. Pook, Naval Instructor. Began trade 1858 as brass and iron finisher. Health is good.

BENJAMIN McCATHRAN
40 years
Machinist A. R.

Born Feb. 1, 1855, in the city of Washington. Has been continuously employed since June 20, 1880. Commended for faithful service during period of employment.

Navy yard workers honored in 1920. Employees commonly joined the workforce as apprentices, learned ordnance manufacturing skills, and retired after decades of loyal and productive service for the Navy.

Repair, the Navy pressed Congress to build a new model basin in the Washington area. In 1937, Secretary of the Navy Claude Swanson, with President Roosevelt's hearty endorsement, named the Navy's newly constructed hull-testing facility in honor of the Navy's hydrodynamic genius. In the years following construction of the David Taylor Model Basin in 1939, an increasing number of operations were shifted from the Washington Navy Yard to the new Carderock, Maryland facility.

These interwar years also saw the United States Navy Band become the service's premier musical organization. During World War I, the Navy Department had combined a 16-piece group from battleship *Kansas* with a 17-piece unit serving presidential yacht *Mayflower* to form the "Washington Navy Yard Band."[38] When President Warren G. Harding traveled throughout the West and to Alaska in 1923, he was accompanied by thirty-five Navy bandsmen. When a sudden heart attack felled Harding in San Francisco, the group had the sad duty of performing "Nearer My God to Thee" as the President's body was placed on board a train for transportation to Washington. Harding's successor, Calvin Coolidge, was so impressed with the quality of the band's music that he readily signed a special act of Congress in 1925 which stipulated that "hereafter the band now stationed at the Navy Yard, known as the Navy Yard Band, shall be designated as the United States Navy Band."[39] During the 1920s the Navy Band began its traditional yearly world tours to represent the U.S. Navy.

The navy yard continued to serve during the interwar years as the U.S. Navy's preeminent ceremonial site in the Washington area. On 11 June 1927, light cruiser *Memphis* arrived at the yard carrying Colonel Charles A. Lindberg, the first person to fly a plane non-stop across the Atlantic Ocean from the United States to France. President Coolidge had insisted that the American hero be returned home in a U.S. warship. More than 9,000 admirers crowded

Above. Women were dedicated employees of the navy yard from the earliest days of its existence. Mrs. Almira V. Brown, still an employee at age 81 in 1920, joined the workforce during the Civil War.

Right. Throughout its history, the navy yard acted as part of the Washington, D.C. community. In the years after World War I, the command sponsored Camp Good Will and other summer camps for children of employees and local indigent families.

Above Left. Navy yard employees worked and often played together. Commandant J. J. Raby threw out the first ball of the 1920 baseball season.

Above Right. The navy yard baseball team scores a success.

Left. Impressed with the quality of music played by the Navy Yard Band, in 1925 Congress passed and President Calvin Coolidge signed a special act designating the group as the United States Navy Band. In this view, the band performs at a navy yard concert during the late 1920s.

into the navy yard between the main gate and the waterfront to catch a glimpse of "Lucky Lindy." In their enthusiasm to get close to the pilot of the "Spirit of St. Louis," fans, reporters, and photographers surged past Marine guards and temporary barriers in a mad rush. Not until Lindberg's car had passed through Latrobe Gate was order once again established.

The Washington Naval Treaty had curtailed construction of new battleships and their heavy ordnance, but the gun factory kept its workers busy during the late 1920s manufacturing new 5-inch and 8-inch guns for cruisers and reconditioning existing 14-inch and 16-inch guns on the old battleships. The labor force totaled more than 4,000 workers in 1929. Despite the London Naval Treaty of 1930 mandating warship reductions, and President Herbert Hoover's decision to drastically cut naval construction and other government expenditures to stave off an economic depression, the gun factory continued to turn out ordnance in the early 1930s.

Soon after Franklin D. Roosevelt was elected President in 1932, he directed the Navy to operate a new presidential yacht from the Washington Navy Yard. For that purpose, in 1933 the Navy acquired the relatively new wooden yacht *Sequoia*. In 1936, *Potomac* took on that responsibility and *Sequoia* became the "barge" for the Secretary of the Navy. For many years afterward, both vessels were common sights on the navy yard waterfront.[40]

Above. *The gun factory designed and tested equipment for the new aviation arm. In June 1923, Langley (CV 1), the Navy's first aircraft carrier, tied up at the navy yard.*

Right. *Charles A. Lindberg, America's hero after his daring and unprecedented solo flight across the Atlantic, disembarks from cruiser Memphis (CL 13) at the navy yard in June 1927.*

The Presidential Yachts

For more than fifty years, the Washington Navy Yard was the home port for yachts and other vessels serving the diplomatic and social needs of the President of the United States.

Despatch and *Dolphin*, while not formally designated presidential yachts, were the first U.S. naval vessels to be employed in "Presidential service" when assigned to the navy yard. In the days before radio communication these fast steamers transported messages between shore-based headquarters and the fleet. *Despatch*, the former merchant steamer *America* purchased by the Navy in 1873, hosted the President, the Secretary of the Navy, and other high-ranking government officials in their visits to the fleet for naval reviews and short-term cruises on the Potomac River from 1880 to 1891. When commissioned in the mid to late 1880s, despatch vessel *Dolphin* was one of the first steel vessels built for the modern Navy. From 1885 to World War I, *Dolphin* routinely carried out her message-delivery responsibility, but on a number of occasions embarked the President. In 1897, for instance, the ship transported President William McKinley to New York City for a ceremony at Ulysses S. Grant's tomb. During the late 19th century, these vessels provided the President with a means for escaping temporarily the pressures of his office and the hot, humid Washington summers.

Sylph (PY 12), purchased in 1898, was the first naval vessel assigned the primary duties of carrying the chief executive of the United States to and from official functions and hosting his social outings. Presidents McKinley, Theodore Roosevelt, and William Howard Taft spent many leisurely hours on board *Sylph* while she was anchored off the coasts of New York and New England. Traditionally, when the President was assigned a new yacht, the old one was handed down to the Secretary of the Navy, so *Sylph* served as the Secretary's personal conveyance, from 1905 to 1929, after *Mayflower* became the presidential yacht.

Mayflower (PY 1), the largest, grandest, and longest-serving of the presidential yachts, was built in Scotland in 1896. Almost the size of a World War II destroyer escort, the yacht boasted a crew of 171 men, including bandsmen and Marine guards. Commissioned originally as a gunboat for the Spanish-American War, *Mayflower* operated off Cuba and transported Admiral of the Fleet George Dewey and his staff.

During the pre-World War I era, the presidential yacht became a "prop of power." U.S. Presidents, like their European counterparts, used luxury yachts to symbolize their nation's military and industrial power and their personal stature. In that regard, *Mayflower* served as the venue for the 1905 diplomatic negotiations sponsored by President Theodore Roosevelt to end the Russo-Japanese War. Roosevelt, who later received a Nobel Peace Prize for this successful effort, introduced Russian and Japanese dignitaries to one another on her decks, and they discussed key issues in her compartments.

The presidential yacht continued to serve the social needs of the nation's principal elected leader. The widower Woodrow Wilson found *Mayflower* an ideal setting for courting Edith Bolling Galt, who became his second wife. He liked *Mayflower's* victorian ambiance and he could find refuge on board from inquisitive Washington reporters. Calvin Coolidge frequently used *Mayflower* to entertain his friends and the press corps. The taciturn and frugal New Englander made sure, however, that the press corps did not have second helpings at the buffet table! With the onset of the Great Depression in 1929, President Herbert Hoover ordered *Mayflower* and *Sylph* decommissioned as economy measures.

Sequoia (AG 23), unlike the yachts before and after her, was not a blue-water, oceangoing vessel. She was built in 1925 as a civilian cabin cruiser for inland waterway operations. The Navy took charge of her when President Hoover indicated he had official and social needs for the vessel—he especially liked to fish. When Franklin D. Roosevelt requested a larger, oceangoing vessel for his use in 1936, the Navy transferred *Sequoia* to the Secretary of the Navy. The vessel operated in support of the Navy's head civilian for the next four decades.

When commissioned in 1936, *Potomac* (AG 25), a former Coast Guard cutter, became the new presidential yacht. The President and

Mayflower, with the President's pendant hoisted, during a 1912 fleet review.

Mrs. Roosevelt embarked on several occasions for cruises in the Chesapeake Bay and to Florida and the Bahamas. In 1939, the President and First Lady hosted the King and Queen of Great Britain on board the vessel during a short day-trip to George Washington's Mount Vernon estate. In 1941, President Roosevelt embarked in *Potomac* for a voyage to Cape Cod, Massachusetts, where he secretly transferred to cruiser *Augusta*, which carried him to Argentia Bay, Nova Scotia, for a meeting with British Prime Minister Winston Churchill. The two leaders reached agreement on the historic Atlantic Charter that defined Allied goals in World War II.

To provide the President with a safer wartime retreat, the Navy established what Roosevelt called Shangri La, now known as Camp David, in the Catoctin Mountains of Maryland. Initially, the commanding officer of the presidential yacht was responsible for the operation of Camp David.

After World War II and Roosevelt's death, the Navy selected another yacht for President Harry S. Truman. *Williamsburg* (AGC 369), which was built in 1931 and had served as a gunboat and a flagship during World War II, transported the President to his Key West, Florida summer White House, to Philadelphia for the annual Army-Navy game, and to other East Coast destinations. The vessel also hosted disabled veterans taking excursions to Mount Vernon. President Dwight D. Eisenhower used *Williamsburg* just once and in 1954 ordered the ship decommissioned as an economy measure.

President John F. Kennedy made use of the Secretary of the Navy's yacht *Sequoia* when he needed to entertain important guests, as he did with Pakistan's President Ayub Khan in 1961. Presidents Richard Nixon and Gerald Ford conducted social activities and hosted foreign dignitaries on board *Sequoia*, but they were the last chief executives to do so. In 1977, President Jimmy Carter, determined to rid the government of ostentation, ordered the yacht sold. Many Americans shared his view, so there was little opposition to the action. In another sense selling the vessel was appropriate, since in the modern era of long-range aircraft, nuclear weapons, and intercontinental ballistic missiles, presidential yachts no longer symbolized for most Americans the power and majesty of the United States. For much of the 20th century, however, the Washington Navy Yard has witnessed history through its visitors, as presidents, emperors, kings and queens, and other world leaders passed through Latrobe Gate to embark in the presidential yachts on the Anacostia River.

Jack A. Green

Of greater significance was Roosevelt's decision soon after taking office to stimulate national recovery from the Great Depression by increasing naval construction. He was also concerned that Congress had failed to maintain even the low number of capital ships and guns allowed the United States in the naval limitation treaties. He persuaded Congress to pass the Vinson-Trammell Act, which provided for warship construction in the categories regulated by the treaties. During the next several years, the Naval Gun Factory produced an impressive quantity of new weapons and ordnance equipment. The workforce, numbering almost 8,000 by 1934, produced new naval rifles ranging from 5-inches to 16-inches, sights, gun directors, and torpedo tubes. The expiration of the London Naval Treaty in 1936 and the increasingly aggressive actions of Germany and Italy in Europe and Japan in the Far East stimulated Congress to approve even more construction for what would become a "two ocean navy." Shipyards began building new battleships, destroyers, and submarines. In May 1939, Congress authorized production of 3,000 aircraft and a 20 percent increase in the warship strength of the Navy.

Above. *Washington resident Thomas King served his country as a "Buffalo Soldier" in the famed 9th U.S. Cavalry Regiment and in the years after World War I as a loyal employee of the Washington Navy Yard. Courtesy of Carl Cole*

Left. *Throughout its history, the navy yard has suffered from flooding of the Anacostia River. Surging waters inundated buildings close to the waterfront during the flood of 1936.*

President and Mrs. Herbert Hoover converse with Commander Louis J. Gulliver, Commanding Officer of Constitution, *when the Navy's oldest commissioned warship visited the navy yard in November 1931.*

That year, as Europe edged toward war, the navy yard manufactured 383 naval rifles of all calibers and reconditioned another 764. Various shops produced over 18,000 spare parts and almost 6,000 optical pieces.

The Washington Navy Yard provided a setting for the meeting of leaders who would lead their nations in the second global conflict of the 20th century. On 9 June 1939, the President and Mrs. Roosevelt welcomed Their Highnesses, the King and Queen of Great Britain, on board presidential yacht *Potomac*, moored at the yard. More than 5,000 invited guests were on hand to witness the event and take part in formal ceremonies. *Potomac*, flying the flag of the President of the United States and the banner of His Majesty, the King of Great Britain, got under way just after noon and shaped a course for President George Washington's Mount Vernon estate. As the vessel headed down the Anacostia River for the afternoon's excursion, the Washington Navy Yard ceremonial battery fired a 21-gun salute and the crowd cheered. The embarked leaders would enjoy few such moments of peace in the months and years ahead.

Left. *The yacht* Potomac *comes alongside the navy yard dock with the President and Mrs. Roosevelt and the King and Queen of Great Britain embarked during the royal visit to America in June 1939.*

Below. *First Lady Eleanor Roosevelt, Queen Elizabeth, King George VI, President Franklin D. Roosevelt, and his Naval Aide, Captain Daniel J. Callaghan (who received a posthumous Medal of Honor for extraordinary valor in World War II), appear at the rail of* Potomac.

Ordnance Nerve Center for a Global Conflict

Soon after the German army invaded Poland on 1 September 1939, sparking the outbreak of World War II, the Navy took action to prepare for the war that President Roosevelt and many naval leaders felt the United States could not avoid. By then, the Naval Gun Factory had provided the Navy with the "finest ship borne artillery of any navy in the world."[41] To maintain that edge, the Bureau of Ordnance took steps to increase ordnance design, testing, and production by the gun factory and hundreds of industrial facilities around the country.

Arsenal of democracy. A skilled worker, one of 26,000 men and women who toiled at the gun factory at the peak of World War II, clears metal cuttings from the barrel of a weapon being machine-rifled for accuracy.

Naval leaders recognized early that only private industry could produce the enormous number of weapons and ordnance systems needed by the fleet to fight a two-ocean, global conflict, but they believed the Naval Gun Factory should continue to be the "nerve center"[42] of the Navy's ordnance design and testing program. The Washington facility was responsible for furnishing blueprints, drawings, and specifications to the nation's industrial plants to ensure ordnance standardization. It also concentrated on the repair of battle-damaged equipment and the production of critical or specialized items, replacement parts and equipment, and prototype ordnance. For instance, the gun factory designed the huge turrets for the 16-inch guns of the *North Carolina-* and *Iowa-* class battleships, new 12-inch, 8-inch, and 6-inch guns that employed the latest in automatic loading and firing devices, submersible deck guns for submarines, and rapid-fire rocket launchers. The yard also was a site for the design and testing of air-to-surface rockets and other aviation ordnance, for which the Navy constructed Building 220 on a high-priority basis. Only three months after the Bureau of Ordnance authorized full-scale production of the technologically advanced variable time (VT) fuse for gun ammunition in 1943, the facility manufactured 2,200 fuses each day, and by June 1945 that figure had increased ninefold.[43]

The facilities at the Washington site employed the largest number of workers of any Navy shore establishment, with the exception of the major navy yards. In mid-1940 there were 10,000 employees at the

Gun factory employees work metal into shape for one of the weapons the U.S. Navy will use in World War II.

yard. By the end of 1944 the figure had risen to 26,000, the wartime peak.[44]

Many of these employees were women, tens of thousands of whom went to work at defense plants across the nation. The women at the Washington Navy Yard, like the men, had to endure long hours of labor in the routinely hot or cold, noisy, and dirty industrial facilities. The women did not share equally in pay and other compensation. The take-home pay for some women was $23 a week while some skilled men received $22 a day! One former female worker observed that the navy yard women

Ordnance Nerve Center for a Global Conflict   71

Above. Work is in full swing at the breech mechanism shop (Building 76) during the war.

Right. A 3-inch .50-caliber, rapid-fire, twin-mount antiaircraft gun after assembly at the gun factory. These weapons were essential to the fleet's defense against Japanese kamikaze attacks in the Pacific.

"would stand for practically anything—five months without sleeping in a bed, a solid year on the graveyard shift so as to be home with the kids during the day, the double job, indigestible lunches, long hours and no promise of a future after the war—all for miserably low wages."[45] The women were partially motivated by the need to supplement their husband's meager military pay. But, they also recognized that their sweat and toil would help their country survive the onslaught by Germany and Japan and perhaps bring about a postwar era of peace and prosperity.

More land to the west of the navy yard was acquired in 1942 to handle the prodigious production of the gun factory and the storage of massive amounts of steel and other materials brought into the site by train and truck. By the end of the war, the Washington Navy Yard administered 132 buildings on 127 acres of land in southeast Washington. In addition to the ordnance design and production facili-

Above. Much like John Dahlgren's men in the experimental battery, these World War II sailors, operating from the top floor of the highest building (157) in the yard, test gun rangefinders by sighting visible landmarks across the Anacostia River.

Left. Sparks and flames leap from cauldrons of molten metal as the gun factory works to arm the Navy for the global conflict.

Sailors from the receiving station unit at the navy yard take part in a ceremony as they prepare for their assignments to the fleet.

ties, the yard hosted a laboratory for investigating mines and certain aspects of ammunition development. After the war the Naval Ordnance Laboratory was moved from the yard into new facilities at White Oak, Maryland.

The navy yard's Medical Department concentrated on reducing on-the-job injuries and illnesses, registering considerable success during the war. Employment of safety engineers and other measures helped reduce the incidence of time lost to illness or injury by more than half from 1942 to 1945. These employees saw to it that workers wore safety shoes, goggles, and protective glasses.

The navy yard also hosted a number of schools, including the Navy Yard Apprentice School and six ordnance and gunnery schools—Gunners Mates and Electric Hydraulic School, Mine Disposal School, Gunnery School, Ordnance School, Optical Primary School, and Advanced Fire-Control School. Between September 1943 and July 1945, the number of officers and men attending these schools rose from 1,658 to 5,822.[46] Many members of the Women Accepted for Volunteer Emergency Service (WAVES) replaced male sailors at navy yard facilities who were then available for service in the combat theaters.

In celebration of Victory in Europe (VE) Day, 8 May 1945, the Navy broadcast to all yard employees President Truman's speech to the nation announcing the German defeat. To celebrate the event, the commandant opened the yard to 5,000 citizens, allowed visitors to tour destroyer *George K. MacKenzie*, and sponsored various programs throughout the facility[47]

Above. *Navy yard employees gather in Leutze Park for entertainment by the United States Navy Band and a well-deserved break from wartime labors.*

Left. *Women shift workers take a meal break during their workday at the gun factory. Thousands of women, many with children at home and husbands far away in combat theaters, worked long, hard hours and for low pay during the war.*

The workers of the Naval Gun Factory could be justly proud of their contribution to victory over the Axis Powers in World War II. During the conflict, the facility produced a prodigious amount of first-rate ordnance for the fleet, including 236 16-inch, 370 12-inch and 8-inch, 134 6-inch, and 427 5-inch guns, plus thousands of turrets, catapults, and antenna mounts.[48]

As a symbol of the factory's importance to the Allied victory, a bronze plaque manufactured by the Naval Gun Factory was affixed to the spot on the deck of battleship *Missouri* (BB 63) where Japanese officials signed surrender documents in Tokyo Bay.[49]

A New Mission for the Navy Yard

On 1 December 1945, the Secretary of the Navy formally changed the name of the Washington Navy Yard to the United States Naval Gun Factory. This recognition was belated, for the yard had been functioning as an ordnance research, design, testing, and production facility since 1886 rather than as a traditional navy yard.⁵⁰ The decision was not well received by some old-time yard workers and Washington residents, who appreciated the location's unique past and its importance to the history of the nation's capital and the Navy.

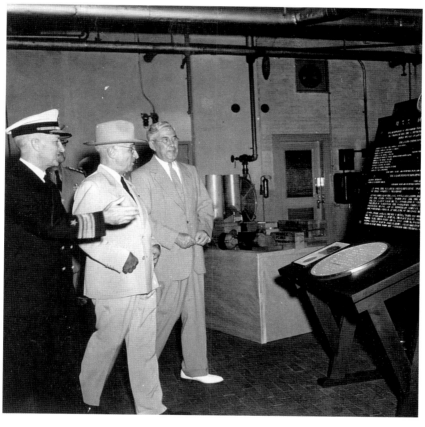

As a fitting symbol of the gun factory's contribution to Allied victory in World War II, President Harry S. Truman views a plaque designed and cast at the yard to commemorate the Japanese surrender on board Missouri.

The post–World War II years saw drastic reduction in ordnance work at the navy yard. With a huge fleet of heavily armed warships already built, the Navy did not need to produce additional guns. Furthermore, during the 1950s the Truman and Eisenhower administrations, hoping to hold down the cost of maintaining large, expensive conventional forces, limited construction of new warships. The Navy added to gun factory woes by stressing the preeminence of the aircraft carrier and the guided missile in a new era of naval warfare. Other naval facilities around the country were better suited than the navy yard to the design, testing, and production of aircraft and radar-guided missile systems. Throughout the early Cold War years, the gun factory laid off an increasing number of employees as the Navy issued fewer and fewer ordnance work orders.

For those workers still on the payroll, there was still plenty to do. In October 1949 the Washington Navy Yard celebrated its 150th anniversary with a week of events that included a Marine Parade and Water Pageant, Power Boat Regatta, a canoe and rowing competition, and an aircraft flyover. President Truman attended several of the events. Unfortunately, the good feeling from these activities was somewhat dissipated in November and December of that year when two passenger planes crashed into the Anacostia River nearby, killing a number of people on board. Navy yard workers and heavy equipment helped raise the aircraft so government investigators could determine the cause of the dual tragedies. The gun factory also continued to maintain the U.S. Senate's subway system, whose two cars were built at the navy yard in 1912.⁵¹

76 The Washington Navy Yard

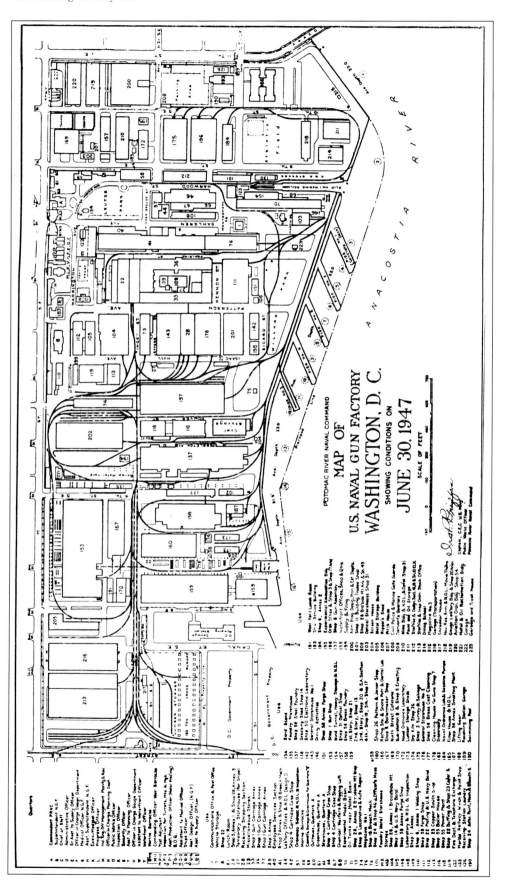

Above. The navy yard's Naval School of Diving and Salvage, operating from barge Tom O'Malley (YRST 5) and buildings ashore, prepared thousands of sailors for hazardous duty with the fleet.

Right. Submarine torpedo tubes and other ordnance produced by the gun factory were heavily used by the fleet in the Cold War era.

Above. *In the years after World War II, the yard continued to serve as a ceremonial site for the Navy and the nation. On 14 April 1947, Rear Admiral Richard E. Byrd (standing behind the microphones)* returned to the United States from a naval exercise in Antarctica. Welcoming the famous explorer home are (left to right) Chief of Naval Operations Fleet Admiral Chester W. Nimitz, Secretary of the Navy James Forrestal, and Rear Admiral Richard H. Cruzen.

Left. *Navy yard workers operate data processing equipment in a gun factory administrative office. With the Navy emphasizing aircraft and missiles for its Cold War missions, lessening the need for guns, the navy yard became more of an administrative center.*

One of the navy yard's responsibilities, maintaining and operating yachts for the President and the Secretary of the Navy, continued well into the Cold War years. In November 1945, the steel-hulled, diesel-powered yacht *Williamsburg* replaced *Potomac* as the presidential yacht. *Williamsburg* served both Truman and Eisenhower who entertained American and foreign dignitaries on board or used the vessel to transport them on visits to the fleet or East Coast ports. In 1953, Eisenhower directed decommissioning of the ship as a cost-cutting measure.[52]

At the height of the Cold War, employees prepared for the worst—a Soviet nuclear attack on Washington, D.C.! Initially, employees were instructed to remain at their workstations and cover their mouths with handkerchiefs to ward off the ill effects of atomic radiation. Soon, however, more serious measures were taken to protect yard personnel. Throughout the 1950s, air raid sirens alerted nineteen auxiliary damage control teams and sent as many as 11,000 workers scurrying into assigned shelters. Planners then expected that the radiation, heat, and blast of a nuclear detonation would cause widespread devastation and kill several thousand workers. Thankfully, such an attack never happened.[53]

In May 1958, the navy yard hosted the return to the United States of the unknown dead from World War II and the Korean War, as it had for the unknown men of World War I years earlier. On the 28th, in a formal ceremony widely reported by the American news media, destroyer *Blandy* carried the remains of the unknown servicemen to the yard, where they were transferred to vehicles that would bring them to Arlington National Cemetery for interment.

With ordnance work assigned to other naval facilities and private industry, the Navy reestablished the Washington Navy Yard as an administrative, supply, and ceremonial center. As it had for many years, the yard served as the "ceremonial quarterdeck" for Washington-area naval commands. The Marine Corps' Ceremonial Guard Company, billeted in Building 58 adjacent to Leutze Park from 1957 to 1977, performed their world-famous precision drills

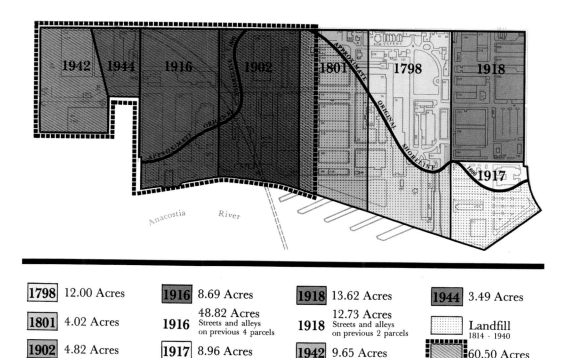

Land acquired by the Washington Navy Yard.

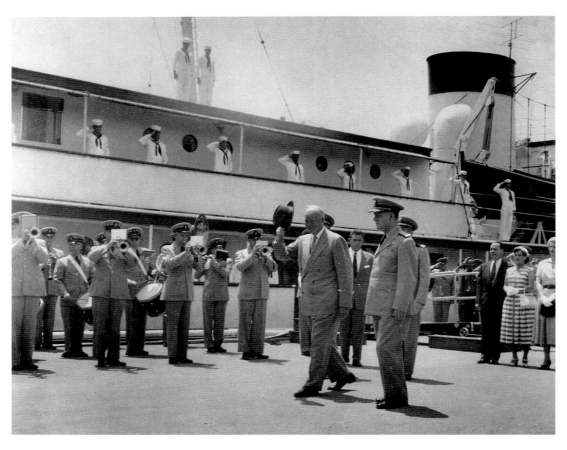

Left. *As it had in the first half of the 20th century, the navy yard continued to maintain the presidential yacht during much of the Cold War. Shortly after President Dwight D. Eisenhower disembarked from* Mayflower *in May 1953, the retired general discontinued operation of the vessel.*

Below. *Destroyer* Blandy *(DD 943), which returned remains of unknown American servicemen to the United States in 1958.*

Above. *Guns manufactured at the navy yard were used against the enemy in every one of the nation's 20th-century wars, including this 16-inch gun on board Wisconsin (BB 64) firing at targets in Iraqi-occupied Kuwait during Operation Desert Storm in 1991.*

Right. *Washingtonians enjoy a "Watergate Concert" at the foot of Memorial Bridge. The gun factory fabricated the band shell and barge combination for the concerts held here during the 1950s and 1960s.*

at the yard and the nearby barracks on 8th and I Streets. In any given year, the navy yard hosted numerous retirements, award presentations, changes of command, and welcoming ceremonies for domestic and foreign dignitaries. In 1959, for instance, U.S. naval leaders welcomed to America naval flag officers from Colombia, Mexico, Greece, and seven other countries. In June 1960, the Navy hosted a formal ceremony in Leutze Park to mark the 100th anniversary of the first visit to the United States by a Japanese ambassador. The commandant presented the crew of Japanese merchant ship *Nippon Maru* with an anchor forged by the plant and inscribed with the words "steadfast in friendship."[54]

The Bureau of Ordnance began to identify missions other than production of heavy ordnance for the Naval Gun Factory. Its shops investigated the use of electronics and plastics in weapons, tested equipment for a satellite navigation system, and produced launchers for the Navy's Tartar surface-to-air missile.[55] In 1959, the Navy changed the name of the United States Naval Gun Factory to United States Naval Weapons Plant, "in recognition of the fact that gun technology was no longer paramount in modern naval combat." By the end of 1961, the plant had completed all of its scheduled ordnance production. In short, "a 114-year tradition of ordnance manufacture on the banks of the Anacostia River was over."[56] In the next two years, the weapons plant was disestablished and half of the buildings and land formerly constituting the navy yard transferred to the General Services Administration.

The United States Navy Band, headquartered at the Washington Navy Yard in the Sail Loft (Building 112) since its establishment, continued to attract

Above. *Admiral Arleigh Burke, former Chief of Naval Operations, shares a laugh with fellow flag officers at a navy yard change-of-command ceremony in which Admiral David L. McDonald (right) succeeded George W. Anderson as CNO. A strong advocate of U.S. naval history, Burke directed establishment of what became The Navy Museum at the yard.*

Left. *A sailor looks over the Fighting Top of USS* Constitution *exhibited in the Navy Museum.*

Right. *Guns and missiles on display in Willard Park.*

Below. *The Navy Museum.*

national attention for the excellence of its programs and accomplished musicians. The loss of nineteen band members in an air disaster in Brazil during February 1960 saddened Americans but did not end the band's well-regarded foreign tour. On many occasions, the national media highlighted the virtuosity of Navy Band members like Master Chief Musician Frank J. Scimonelli, at one time the world's foremost English post horn player. The band routinely performed at presidential inaugurals, Washington memorial dedications, U.S. Capitol concerts, and funeral services at Arlington National Cemetery for sailors who made the ultimate sacrifice in World War II, Korea, and Vietnam. The United States Navy Band also entertained the public at "Watergate Concerts," for many years held on board a barge manufactured by the Naval Gun Factory and moored at the foot of Memorial Bridge.[57]

Recognizing that the Washington Navy Yard was the Navy's oldest and one of its most historic sites, in 1961 Chief of Naval Operations Admiral Arleigh Burke persuaded Secretary of the Navy John B. Connally to establish the U.S. Naval Historical Display Center (later the U.S. Navy Memorial Museum, then The Navy Museum) in Building 76, the old breech mechanism shop. The purpose of the museum's collections was to "inspire, inform, and educate service personnel and the public in naval history, traditions, heritage and scientific contribution."[58]

The cavernous building eventually displayed the fighting top of frigate *Constitution*, World War II aircraft and weapons, the *Trieste* deep-diving vessel, and thousands of other unique artifacts, paintings, ship models, medals, flags, and uniforms. Exhibits cover all periods in U.S. naval history and interpret the service's role in war and peace, diplomacy, exploration, and scientific research.

Almost from its beginning, the yard has displayed at outdoor sites larger artifacts, including bronze and iron guns from the Navy's past wars and skirmishes.

84   The Washington Navy Yard

Above. *The float built by the navy yard and carrying a PT boat (actually PT 796) moves along Pennsylvania Avenue during President John F. Kennedy's inaugural parade in January 1961.*

Left. *President Kennedy speaks on board U.S. Coast Guard sail training ship* Eagle *at the navy yard.*

The majority of outdoor artifacts are found in Willard Park, named for a former commandant. The historic pieces in this site across from the museum include U.S. and captured Confederate cannon from the Civil War, a 16-inch naval rifle, a 14-inch railway battery, a Talos surface-to-air missile, and a Vietnam-era fast patrol craft, or "Swift boat."

The Navy established another exhibit center at the yard, the Combat Art Gallery (later the Navy Art Gallery). The gallery exhibits oils and watercolors by Navy and civilian artists, including Thomas Hart Benton, Reginald Marsh, and others, who depicted sailors and marines taking part in World War II, the Korean and Vietnam wars, and in other operations of the Cold War era.

In 1977, *Barry*, a *Forrest Sherman*-class destroyer that served for twenty-six years with the Atlantic and Pacific fleets and saw action in the Vietnam War, was permanently moored at Pier 2 on the waterfront and opened for public visitation.

The navy yard continued to serve as the venue for welcoming foreign heads of state and naval leaders to the United States and as an appropriate mooring place for visiting naval combatants, training vessels, and oceanographic research ships. In any given year, numerous U.S., British, Canadian, and other warships and non-combatant vessels put in at the Washington Navy Yard. For instance, in early 1961, motor torpedo boat *PT 796* arrived at the yard for the inauguration of John F. Kennedy, whose exploits as a PT boat skipper in World War II were well known. For the inaugural parade, the navy yard fabricated a cradle and float to hold the boat, temporarily designated *PT 109*. That summer, a party led by President Kennedy and President Mohammad Ayub Khan of Pakistan boarded the Secretary of the Navy's yacht *Sequoia*, the privately owned vessels *Honey Fitz* and *Patrick J*, and naval vessel *Guardian*, which carried the group to Mount Vernon for a state function.[59]

Above. *The Navy Art Gallery (formerly the Combat Art Gallery).*

Left. *"Score Another for the Subs" by Thomas Hart Benton (Navy Art Collection).*

The Navy Museum

The Navy Museum opened to the public in 1963 to collect, preserve, display, and interpret artifacts and artwork for the general public. Called one of Washington, D.C.'s most "user-friendly" museums, The Navy Museum exhibits an extraordinary collection of ship models, uniforms, medals, ordnance, photographs and fine art. Housed in the former Breech Mechanism Shop of the old Naval Gun Factory, the museum collection includes an F4U Corsair, nicknamed "Big Hog"; a twin mount 5-inch .38 caliber anti-aircraft gun; the foremast Fighting Top from frigate *Constitution;* and the bathyscaphe Trieste that descended nearly seven miles to the deepest location in the Pacific Ocean.

Tracing wars, battles, and crises from the American Revolution through the Vietnam War, The Navy Museum presents over 225 years of naval and maritime history. The museum's most comprehensive exhibit, "In Harm's Way," chronicles the Navy's role in World War II from the attack on Pearl Harbor in 1941 to Japan's surrender in 1945. Its three sections examine the Pacific and Atlantic campaigns and the home front effort. Thematic exhibitions, such as "Underwater Exploration," "Polar Exploration," and "Commodore Matthew Perry and the Opening of Japan," feature the Navy's diplomatic and peacetime contributions. World War II submarine periscopes, gun mounts, and other hands-on objects provide visitors with an interactive museum experience.

The museum offers the surrounding community an active education and public programs schedule. Teachers are encouraged to sign up for curriculum-based school programs and tours for grades one through twelve. Tours for senior citizens, families, naval reunion groups, and general audiences are also available. The Navy Museum sponsors evening events throughout the year including lectures, slide presentations, book-signings, and concerts performed by the United States Navy Band and traditional maritime musicians. Each fall, the museum hosts the Seafaring Celebration, a family festival highlighting naval and maritime traditions. Activity workshops, demonstrations, storytelling, and musical performances provide thousands of visitors with exciting educational experiences throughout the day. All of the museum's programs are open to the public and free of charge.

To increase public outreach, The Navy Museum establishes partnerships with other educational organizations. The partnership between The Navy Museum and the Alexandria Seaport Foundation is especially productive. Begun in March 1998, it has developed into a cooperative community boat-building project that serves youngsters from the District of Columbia and surrounding communities. The Navy Museum, whose mission is to increase outreach to the public, was attracted by the Alexandria Seaport Foundation's success with past programs that taught the value of teamwork, in addition to basic boat-building skills. Both organizations are dedicated to the education and inspiration of local youth.

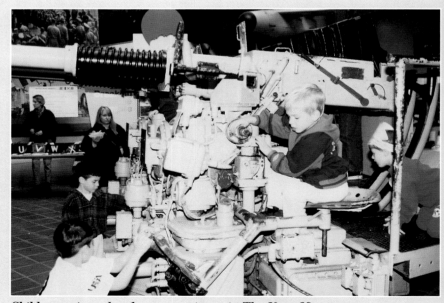
Children enjoy a hands-on experience in The Navy Museum.

First Lady Hillary Rodham Clinton visits The Navy Museum in March 1999 to lend her support to preserving America's historic treasures. Among others, she is accompanied by *(left to right)* Anthony Williams, Mayor of the District of Columbia; Rear Admiral Arthur N. Langston, Director Navy Staff; Congresswoman Eleanor Holmes Norton, (D–D.C.) and William S. Dudley, Director of Naval History *(far right)*.

The first boat-building program at The Navy Museum began in the summer of 1998 with high school students from the Greater Washington Urban League's Environmental WALL program. The young people built two, 12-foot rowing skiffs for use in taking water samples from the Anacostia and Potomac Rivers. After the boats were completed, the museum held a "Blessing of the Fleet" ceremony to honor the boat builders and their finely made craft. On 5 March 1999, First Lady Hillary Rodham Clinton officiated at a similar "Blessing of the Fleet" ceremony held in The Navy Museum. Honored guests included Secretary of the Navy Richard Danzig, Congresswoman Eleanor Holmes Norton (D–D.C.), and District of Columbia Mayor Anthony Williams.

Priceless artifacts, didactic exhibits, and diverse public programs attract nearly 350,000 people annually to The Navy Museum. Each visitor leaves the museum with an appreciation of America's rich naval and maritime heritage.

Susan Scott

Fast attack craft continued to be popular attractions during the Kennedy administration. In May 1963, Lieutenant (jg) John R. Graham brought the vessel under his command, Norwegian-built *PTF 3*, from Norfolk to the navy yard. Two days later, Secretary of the Navy Fred H. Korth boarded the vessel for a demonstration in the Potomac River of the vessel's speed, maneuverability, and other characteristics. Impressed with the fast patrol boat's potential for use in special operations, Korth, other naval leaders, and Secretary of Defense Robert S. McNamara purchased more boats in the class from Norway and deployed them to Southeast Asia. *PTF 3* and her sisters of the "Nasty" class were involved in raids on North Vietnam's coast that helped spark the Tonkin Gulf Incident of August 1964 and reflected the Navy's and the nation's increasing involvement in the Vietnam War.[60]

Right. *"Queen of the Fleet" by Navy artist Erick Marshall Murray (Navy Art Collection).*

Below. *Captain William L. McGonagle, wearing the Medal of Honor he was awarded at a navy yard ceremony for his courageous leadership as Commanding Officer of Liberty (AGTR 5).*

On 1 January 1965, the Navy created Naval District Washington, headquartered in the Washington Navy Yard, to take over responsibilities formerly handled by the Potomac River Naval Command and the Annapolis-based Severn River Naval Command. Rear Admiral Andrew J. Hill, the first Commandant Naval District Washington, was charged with administering approximately 100 naval activities in the District of Columbia, Annapolis, and the Maryland and Virginia counties of the national capital region.

On 13 July 1966, President Lyndon B. Johnson visited the yard to commission a new research vessel, U.S. Coast Geodetic Survey Ship *Oceanographer*. The President called for a "new age of exploration" to exploit "an unknown world at our doorstep. It is really our last frontier here on earth. I am speaking of mountain chains that are yet to be discovered, of natural resources that are yet to be tapped, of a vast wilderness that is yet to be charted. This is the sea around us."[61]

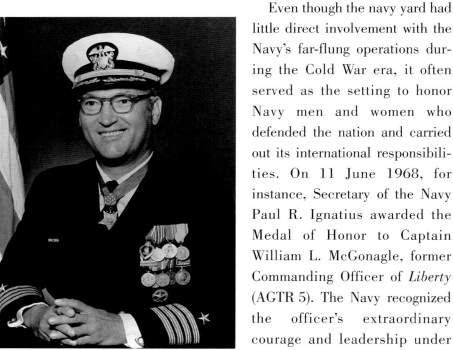

Even though the navy yard had little direct involvement with the Navy's far-flung operations during the Cold War era, it often served as the setting to honor Navy men and women who defended the nation and carried out its international responsibilities. On 11 June 1968, for instance, Secretary of the Navy Paul R. Ignatius awarded the Medal of Honor to Captain William L. McGonagle, former Commanding Officer of *Liberty* (AGTR 5). The Navy recognized the officer's extraordinary courage and leadership under fire when Israeli aircraft and surface vessels attacked his ship in the Mediterranean in 1967. Among the guests of honor at the navy yard ceremony were some surviving members of his crew,

Right. Liberty *after attack by Israeli forces in June 1967.*

Below. *Yacht* Sequoia *(AG 23) off the navy yard in the 1960s.*

the Chairman of the Joint Chiefs of Staff, and the Chief of Naval Operations.

With the demolition of the Main Navy Building on Constitution Avenue in 1970, the Navy leased commercial property in Northern Virginia for some of its Washington-area activities and renovated government-owned buildings at naval facilities for others. As part of this effort, the Navy began consolidating its historical program, then spread around the national capital region, at a most appropriate site, the Washington Navy Yard. Joining the Navy Memorial Museum and the Operational Archives, the latter of which was moved to the navy yard from Arlington, Virginia, in 1963, was the Naval History Division.[62]

The Naval History Division had evolved almost from the beginning of the United States Navy. Its first component was the Navy Department Library. President John Adams established the library in 1800 to provide Secretary of the Navy Benjamin Stoddert with books on "naval architecture, navigation, gun-

The Navy Yard's Historic Ordnance Collection

Captured 18th-century guns in Leutze Park.

Visitors to the Washington Navy Yard notice the array of early bronze and iron artillery displayed in and around Willard and Leutze Parks. These pieces are visible reminders of the navy yard's service as a shipyard, an operating base, a manufacturing plant, and an ordnance test facility.

Trophies of America's wars and expeditions have been brought to the nation's capital since the age of sail. The navy yard collection includes guns used, or captured, by American forces from the Quasi-War with revolutionary France through World War I. A French bronze 4-pounder Gribeauval-pattern cannon, a type that armed Napoleon's field artillery, was cast at Lyons in 1793 and bears the revolutionary motto "liberté, egalité." The oldest piece in the collection, cast in 1686 for King Charles II of Spain, bears the name "San Bruno" given to it by its makers. It honors Saint Bruno, an 11th-century scholar and founder of the Carthusian order of monks, who might have been a bit surprised to find his name on an artillery piece.

A large bronze gun of unusual form, taken from a Japanese coastal fort, recalls the time when Commodore Matthew Perry first opened Japan to American commerce. Feudal clans of southern Japan resented what they saw as foreign intrusion into their country, and fought a civil war for supremacy. Their coastal forts fired on French and Dutch ships and on the American warship *Wyoming*. An international naval squadron retaliated in 1864, silencing the batteries after a heavy gunnery duel. This success is credited with ending the anti-foreign movement in Japan.

Two heavy bronze guns, cast at Barcelona in 1788 and purchased by the Pasha of Tripoli to arm two of his gunboats, commemorate the Barbary Wars. Stephen Decatur captured these gunboats in Tripoli Harbor in 1804, and the frigate *Constitution* carried them back to Washington. Displayed with them are two guns made in 1740 for King Louis XV of France. Early muzzle-loading guns were made in many calibers and dimensions. These guns were made under one of the first rationalized systems of artillery design. Although records of these old guns are sparse at best, these guns are believed to have been purchased in the late 1790s, when war with revolutionary France seemed likely. A number of stubby-barreled howitzers were designed to fire early explosive shells. Two of these howitzers, of unusual form, bear the Lion of Saint Mark on their barrels, showing that they were made at the famed Arsenal of Venice, a combined shipyard and ordnance plant that built and armed Venetian warships during the time when this Italian city-state was a major naval power.

In the 1840s the navy yard took on the task of ordnance development. Lieutenant John Dahlgren built an "experimental battery" on the waterfront and began testing guns and ammunition by firing them down the Anacostia River. The yard's collection reflects these days of experimentation. Two wrought-iron smoothbore guns, made for the Navy in 1844, were designed by Daniel Treadwell, a Massachusetts inventor. They gave satisfaction in tests, but appear to have been too costly for service. After the Civil War many designers worked to develop serviceable breech-loading artillery. Colonel Theodore Yates designed an unusual "clamshell" breech mechanism that worked well with yacht saluting guns but proved unable to handle the powder pressures involved in actual gunnery. A 3.25-inch Yates gun, thought to be the only one of its kind, was apparently tested by the Navy but rejected for service.

A 12-inch shell gun, one of several prototypes developed in the 1840s by Commodore Robert Stockton, holds a conspicuous place in Willard Park. Called the "Mersey Gun," it was made in England to replace Stockton's "Peacemaker," which had exploded in 1844 during demonstration firing on board *Princeton* in the Potomac River. An investigation convinced the Navy that the technology of that time could not yet turn out reliable large-caliber iron guns, so guns of this size did not enter the fleet until the Civil War. Around this gun are grouped five heavy guns designed during that war by Commander John Mercer Brooke, chief of ordnance in the Confederate Navy. His guns were rugged and powerful, and some have judged them to have been the best American-designed guns of their day.

Rifling began to be used in artillery pieces at the time of our Civil War to give them greater power and accuracy. Many early rifled guns were older smoothbores, rifled and usually reinforced to take the higher pressures involved. Five guns across the street from The Navy Museum illustrate the variety of such attempts; American and British guns, converted to rifles, include an American gun made before the War of 1812.

These, along with other guns displayed in the navy yard, are studied with interest by historians and antiquarians. Along with comparable ordnance collections at the Naval Academy and nearby Fort Lesley J. McNair, they serve as a three-dimensional technical archive and a reminder of the efforts and sacrifices made by generations of American soldiers and sailors.

John C. Reilly Jr.

Civil War-era cannon exhibited on Dahlgren Avenue next to The Navy Museum.

nery, hydraulics, hydrostatics, and all branches of mathematics subservient to the profession of the sea" and the "lives of all the admirals . . . who have distinguished themselves by the boldness and success of their navigation, or their gallantry and skill in naval combats."63 During the 19th century, the library was housed in the State, War and Navy Building (now the Executive Office Building next to the White House). In 1882, as the Navy underwent rejuvenation after the lean post-Civil War years, James Russell Soley, a naval officer, respected international lawyer, prolific writer, and Naval Academy professor, was assigned responsibility for managing the library. He improved the library's holdings by gathering rare books, professional journals, official records relating to the Civil War, and prints and photographs of battles and warships. Congress supported his efforts by appropriating an initial $2,640 for the location, collection, and publication of documents on the conflict. Soley continued to oversee the Office of Library and Naval War Records (after 1915 Office of Naval Records and Library) when he was appointed Assistant Secretary of the Navy in 1890. Between 1894 and 1927, Soley and his successors published the 31-volume *Official Records of the Union and Confederate Navies in the War of the Rebellion*.

During World War I, Secretary of the Navy Josephus Daniels directed historical sections in London and Washington to gather documents on the Navy's involvement in that conflict. In 1921, Captain (later Commodore) Dudley Knox, who would become the "driving force of the Navy's historical programs for the next twenty-five years," took the helm of the department's document collection, publication, and library sections. Knox, awarded the Navy Cross for his support of the war effort in London by Admiral William S. Sims,

Above. *Commodore Dudley W. Knox, the Navy's influential and gifted chief historian during the mid-20th century.*

Left. *Assistant Secretary of the Navy James R. Soley encouraged the collection and preservation of books and documents on the history of the U.S. Navy at the end of the 19th century.*

The Dudley Knox Center for Naval History (Building 57).

wartime commander of U.S. naval forces in Europe, certainly had the aptitude for his new assignment. The veteran naval officer had a sharp intellect, gift for writing, and deep knowledge of his navy's history. As a result of Knox's successful actions to organize the Navy's historical materials and his friendship with Franklin D. Roosevelt and other prominent leaders, the Office of Naval Records and Library gained a "national and international reputation in the field of naval archives and history."[64] In 1930 Knox was named Curator for the Navy Department and assigned responsibility for the collection and preservation of the Navy's art objects and artifacts of historical importance. At President Roosevelt's request and with his support, during the 1930s and early 1940s Knox published thirteen volumes documenting the Navy's part in the early 19th century Quasi-War with France and wars with North Africa's Barbary states.

With the outbreak of World War II, Knox established a unique archival system to handle the hundreds of thousands of documents he knew would begin pouring into the Navy Department. He especially wanted to make the collection usable for the researchers and historians who would analyze the Navy's operations and administration in the greatest war of the 20th century. To manage the monumental archival effort, the captain brought on board Naval Reserve officers trained as historians at the nation's finest universities. In addition, Samuel Eliot Morison, a scholar already internationally renowned for his work in naval and maritime history, persuaded Roosevelt to employ him in the preparation of a history of wartime naval operations based on these records.

In 1944, Secretary of the Navy James Forrestal appointed retired Admiral Edward C. Kalbfus as Director of Naval History to oversee the Navy's various historical efforts, including those under Knox.

The Marine Corps Historical Center (Building 58), as seen from Leutze Park.

Another reorganization took place in March 1949, when a new Naval Records and History Division (after 1952 simply the Naval History Division) was established in the Office of the Chief of Naval Operations to manage all of the service's historical programs. Throughout the 1950s, these programs were centered in the Main Navy Building on Constitution Avenue. In 1971, a year after most Naval History Division components had moved to the Washington Navy Yard, it was renamed the Naval Historical Center.

Other new postwar tenants of the Washington Navy Yard included the Chesapeake Division, Naval Facilities Engineering Command, Naval Dental Clinic, Naval Investigative Service, Navy Publications and Printing Service, the Defense Department Computer Institute, Naval Regional Data Automation Command, Weapons Engineering Support Activity, and the Military Sealift Command.[65]

In 1973, the U.S. Congress' Joint Committee on Landmarks identified the Washington Navy Yard "Historic District" as an important landmark because of its significant contribution to the "cultural heritage and visual beauty of the District of Columbia." The National Park Service also listed the yard on the National Register of Historic Places.

Presidential yachts continued to operate at the navy yard. President Richard M. Nixon used *Potomac* often for social functions and occasionally as a convenient venue for thinking through policy decisions. After conferring with his advisors on board the ship in May 1972, Nixon ordered the mining of North Vietnam's harbors. His successor, Gerald Ford, hosted Japanese Emperor Hirohito on board for a cruise down the Potomac. As had Herbert Hoover and Dwight D. Eisenhower, President Jimmy Carter discontinued use of presidential yachts in 1977.[66]

That same year, the Chief of Naval Operations' residence at the Naval Observatory on Massachusetts Avenue was assigned to the Vice President of the United States. In 1978, the Navy chose Tingey House as the new official residence of the CNO. Given the historical importance and ceremonial role of the navy yard and the CNO's close association with those duties, the choice was appropriate. As with most homes that have served generations of families,

Tingey House experienced its share of sadness. On 16 May 1996, Admiral Jeremy Boorda, a dynamic, well-liked and respected Chief of Naval Operations, took his own life on the grounds of the residence.

With the navy yard becoming a popular tourist site in the nation's capital during the 1980s, the Navy presented a weekly "Summer Ceremony" that featured the United States Navy Band and its various components, including the Sea Chanters, Country Current, Commodores, and Port Authority. The U.S. Navy Ceremonial Guard capped the musical performances with precision drills.

Sailors and marines render honors as ceremonial guns boom during an official function in Leutze Park, a common occurrence during the post–World War II years. Behind the formation are Latrobe Gate (left) and Tingey House.

In 1982 the Director of Naval History, retired Rear Admiral John D.H. Kane Jr., concentrated the various branches of the Naval Historical Center in the navy yard's historic precinct around Building 57, fittingly named the Dudley Knox Center for Naval History. Established close by in Building 58 was the Marine Corps Historical Center, whose histories, archives, museum, library, and other branches focus on preserving and interpreting the operational and institutional experience of the Corps. Museum displays of combat art, weapons, uniforms, and other artifacts complement the center histories that detail marine valor and sacrifice in the nation's wars and crises.

The ending of the Cold War opened a new era in U.S.-Soviet relations. As a reflection of the changing

An aerial view of the navy yard in 1998. Visible in the foreground are Latrobe Gate, the rear of Tingey House, and Leutze Park. Decommissioned destroyer Barry (DD 933) is moored on the Anacostia River waterfront.

An aerial view of the Washington Navy Yard looking north. Visible in the foreground are The Navy Museum and Building 1.

times, in December 1988 six Soviet military museum directors toured the Marine Corps Historical Center for briefings on computer applications for artifact cataloging, history writing, and publication printing. The following April, Colonel General Dmitri A. Volkogonov, the Soviet Union's most prominent military historian and biographer of Joseph Stalin, led a delegation that visited the Naval Historical Center. The general expressed the hope that the relationship between the USSR and the United States would improve, "not just historically, but politically as well, because if two such powers can learn to live with each other in a friendly manner, it will be better for the entire world."[67]

The close of the Cold War also had a positive impact on the Washington Navy Yard. Congress began to cut the annual Defense Department budgets in the early 1990s and to close military bases. The Navy's 1990 Master Plan, strongly endorsed by the Base Closure and Realignment Commission (BRAC) in 1993 and again in 1995, called on its Washington-area commands to drastically reduce high-cost commercial leases in Northern Virginia and concentrate their naval activities on government-owned land at the navy yard. As a result, in the mid-1990s, the Navy began converting historic industrial buildings on the western side of the yard for use as office spaces. The Office of the Judge Advocate General, the Naval Facilities Engineering Command, and some staff branches of the Office of the Chief of Naval Operations relocated there. Two parking garages were also built to accommodate the several thousand employees of these activities.

The modernization of historic navy yard buildings and beautification of the surrounding grounds generated positive publicity for the Navy and its relationship to Washington, D.C. But the effort also brought to the surface the environmental degradation caused by almost 200 years of industrial production that polluted the shore site and the Anacostia River. In the summer of 1998, the Environmental Protection Agency named the Washington Navy Yard as a Superfund Site. Efforts were soon begun to determine the extent of the pollution and the requirements for dealing with any contaminants discovered.[68]

Going beyond the BRAC recommendations, the Navy took action in the late 1990s to develop the Washington Navy Yard as the future heart and soul of the naval service in the national capital region. Plans were implemented for the navy yard to serve as the home of major Washington-area commands, including the Naval Sea Systems Command, and the yearly host for thousands of American and foreign tourists. The concept entailed developing a waterfront commercial hotel and conference center, boat marina, and pedestrian amenities, and enhancing museum and outdoor artifact displays. Befitting its historic importance to the Navy and the nation, the Naval District Washington *Naval Station Washington Master Plan* of February 1998 envisioned the Washington Navy Yard as the "Quarterdeck of the Navy representing the best of the Navy and the United States of America" in the 21st century.[69] The Washington Navy Yard, a national and local treasure, thus confidently prepared to begin a new era in U.S. naval history.

An artist's rendering of commercial and government facilities intended to grace the Anacostia waterfront—a vision of the Washington Navy Yard for the 21st century.

Notes

1. Quoted in Peck, *Round-Shot to Rockets*, 6.
2. Bureau of Medicine and Surgery Historian's Office, "Washington, D.C." Narrative Summary, 9 February 1999.
3. Pitch, *The Burning of Washington*, 90.
4. Ibid., 101.
5. Ibid., 102.
6. Peck, *Round-Shot to Rockets*, 62.
7. Canney, *Old Steam Navy*, app. A; Canney, *Lincoln's Navy*, 48.
8. Schneller, *A Quest for Glory*, xiv, 154.
9. Ibid., 74.
10. Ibid., 81
11. Akers, "Freedom without Equality," 26–27.
12. Schneller, *A Quest for Glory*, 152.
13. Ibid., 82.
14. Ibid., 73.
15. Ibid., 180.
16. Dudley, *Going South*, 8.
17. Schneller, *A Quest for Glory*, 183.
18. Akers, "Freedom without Equality," 35.
19. Schneller, *A Quest for Glory*, 185.
20. Ibid., 186.
21. Quoted in Bruce, *Lincoln and the Tools of War*, 17.
22. Quoted in Peck, *Round-Shot to Rockets*, 139.
23. Guttridge, "Identification and Autopsy of John Wilkes Booth," 17.
24. Quoted in Spencer, *Raphael Semmes*, 192.
25. Quoted in Ibid., 195.
26. Sprout and Sprout, *The Rise of American Naval Power*, 195.
27. Quoted in Peck, *Round-Shot to Rockets*, 167.
28. Glasow, "Prelude to a Naval Renaissance," 304.
29. Still, *Ironclad Captains*, 47.
30. "Recreator Year Book: Washington Navy Yard, Nineteen-Twenty," 93, Navy Department Library Rare Book Collection.
31. Carlisle, *Where the Fleet Begins*, 468.
32. Ibid., 109.
33. Society of Naval Architects and Marine Engineers, *Historical Transactions*, app. I, 319.
34. Carlisle, *Where the Fleet Begins*, 60.
35. Ibid., 137.
36. Ibid., 111.
37. Ibid., 132.
38. United States Navy Band, "History of the United States Navy Band," February 1999.
39. Ibid.
40. *Dictionary of American Naval Fighting Ships* 5:63; 6:445.
41. Furer, *Administration of the Navy Department in World War II*, "The Bureau of Ordnance," 341.
42. Peck, *Round-Shot to Rockets*, 234.
43. Gernand, "Event Chronology: Washington Navy Yard, Washington, DC," 38.
44. Bureau of Ordnance, *U.S. Naval Gun Factory*, 250.
45. Anthony, "Working at the Navy Yard," 95.
46. Bureau of Ordnance, *U.S. Naval Gun Factory*, 364.
47. Gernand, "Event Chronology," 39.
48. Ibid.
49. Ibid., 40.
50. The superintendent of the Naval Gun Factory was also made responsible for the Potomac River Naval Command that was established right after Pearl Harbor to manage the Washington region's naval facilities and resources.
51. Gernand, "Event Chronology," 41, 45.
52. *Dictionary of American Naval Fighting Ships* 8:372.
53. Gernand, "Event Chronology," 42–45.
54. U.S. Naval Weapons Plant, Command History, 1960.
55. Gernand, "Event Chronology," 46, 48, 49.
56. Daverede, "From Broadsides to BRAC," 29.
57. United States Navy Band, "History of the United States Navy Band," February 1999.
58. Morgan and Leonhart, *A History of the Naval Historical Center and the Dudley Knox Center for Naval History*, 10.
59. *Dictionary of American Naval Fighting Ships* 6:445; Naval Station Washington, Command History, 1961.
60. Marolda and Fitzgerald, *From Military Assistance to Combat*, 205.
61. *Lyndon B. Johnson, 1966* in *Public Papers of the Presidents*, 722–24.
62. Chesapeake Division, Naval Facilities Engineering Command, *Washington Navy Yard Master Plan, March 1982*, 6.
63. Quoted in Morgan and Leonhart, *History of the Naval Historical Center and the Dudley Knox Center for Naval History*, 1.
64. Ibid., 6.
65. Ibid., 12.
66. *Dictionary of American Naval Fighting Ships* 6:445.
67. Quoted in *Sea Services Weekly*, 4 May 1990, 9; Gernand, "Chronology," 50.
68. *The Washington Post*, 24 July 1998; Naval District Washington, *Naval Station Washington Master Plan*, February 1998, 10–11.
69. *Naval Station Washington Master Plan*, 2.

Bibliography

Archival and Special Collections

"Compilation of Historical Data, 1799–1938." Commandant to Officer in Charge, Office of Naval Records and Library, 4 November 1938. Operational Archives, Naval Historical Center, Washington, DC.

Gernand, Bradley E. "Event Chronology: Washington Navy Yard, Washington, D.C." Naval District Washington Public Affairs Office, 1999.

"History: Navy Yard-Washington, 1799–1921." Original history of the Washington Navy Yard revised and brought up to date. 2 vols. Unpublished manuscript in Special Collection, Navy Department Library, Naval Historical Center.

"Narrative History of the Potomac River Naval Command." 25 October 1945. In *United States Naval Administration in World War II*, Vol. 135. Rare Book Collection, Navy Department Library.

Naval District Washington. Command Histories, 1965–Present. Post-1974, Post-1990 Command Files. Operational Archives, Naval Historical Center.

Office of the Assistant Secretary of Defense–Public Affairs). News Release. "Two River Commands to Be Merged to Form Naval District Washington, D.C." 31 December 1964. Reference File, Navy Department Library.

Potomac River Naval Command. Command Histories, 1959–1965. Post–1946 Command File. Operational Archives, Naval Historical Center.

Rear Admiral David Watson Taylor. Biography. Operational Archives, Naval Historical Center.

"Recreator: Year Book: Washington Navy Yard, Nineteen-Twenty." Rare Book Collection, Navy Department Library.

Reference Files [information regarding the U.S. Marine Corps presence in the Washington Navy Yard]. Reference Branch, Marine Corps Historical Center, Washington, DC.

Shiner, Michael. "Diary 1865–1913." Manuscript Division, Library of Congress, Washington, DC.

U.S. Naval Weapons Plant, Washington, DC. Command Histories, 1959–1962. Post-1946 Command File. Operational Archives, Naval Historical Center.

U.S. Navy. Bureau of Ordnance. *United States Naval Administration in World War II*. Vol. 129, *U.S. Naval Gun Factory*. Washington, 1945. Rare Book Collection, Navy Department Library.

Published Books and Reports

Bruce, Robert V. *Lincoln and the Tools of War*. Indianapolis: The Bobbs-Merrill Co., Inc., 1956.

Bureau of Yards and Docks. *Building the Navy's Bases in World War II: History of the Bureau of Yards and Docks and the Civil Engineer Corps, 1940–1946*. Washington: The Bureau/GPO, 1947.

Canney, Donald L. *Lincoln's Navy: The Ships, Men, and Organization, 1861–1865*. Annapolis: Naval Institute Press, 1998.

———. *The Old Steam Navy*. Vol. 1, *Frigates, Sloops, and Gunboats, 1815–1885*. Annapolis: Naval Institute Press, 1990.

Carlisle, Rodney P. *Where the Fleet Begins: A History of the David Taylor Research Center, 1898–1998*. Washington: Naval Historical Center, 1998.

Chesapeake Division, Naval Facilities Engineering Command. *Washington Navy Yard Master Plan, March 1982*. Washington: CHESDIVNAVFAC, 1982.

Coletta, Paolo E. ed. *United States Navy and Marine Corps Bases, Domestic*. Westport, CT: Greenwood Press, 1985.

Crane and Gorwick Associates, Inc. *Development Plan-Washington Navy Yard/U.S. Naval Station, Washington, D.C*. Technical Report #1, Historic Development. Washington: Crane and Gorwick, 1981.

Dolph, James, and Naval Reserve Detachment 901. *Washington Navy Yard: Building and Structure Summary of Use Database*. Washington: Environmental and Safety Office, NDW, 1997.

Dudley, William S. *Going South: U.S. Navy Officer Resignations & Dismissals on the Eve of the Civil War.* Naval Historical Foundation Publication, series 2, no. 27. Washington: Naval Historical Foundation, 1981.

Engineering Field Activity Chesapeake. Naval Facilities Engineering Command. *Special Study: Cultural Resource Management of Historic Washington Navy Yard Building Drawings.* November 1994.

Farnham, F. E., and J. Mundell. *History and Descriptive Guide of the U.S. Navy Yard, Washington, D.C.* Washington: Gibson Bros., Printers and Bookbinders, 1894.

Furer, Julius Augustus. *Administration of the Navy Department in World War II.* Washington: Naval History Division, 1959.

Hibben, Henry B. *Navy-Yard, Washington. History from Organization, 1799 to Present Date.* Washington: GPO, 1890.

James, Stephen R., Jr. *Underwater Archaeological Investigations: Washington Navy Yard, Anacostia Waterfront, Washington, D.C.* Draft Report prepared by Panamerican Consultants, Inc. Washington: Naval Facilities Engineering Command, October 1994.

Marolda, Edward J., and Oscar P. Fitzgerald. *The United States Navy and the Vietnam Conflict.* Vol. 2, *From Military Assistance to Combat.* Washington: Naval Historical Center, 1986.

Morgan, William James, and Joye E. Leonhart. *A History of the Dudley Knox Center for Naval History.* Washington: Dudley Knox Center for Naval History, 1981.

———. *A History of the Naval Historical Center and the Dudley Knox Center for Naval History.* Washington: Naval Historical Foundation, 1983.

Naval District Washington. *Historic Precinct: Washington Navy Yard.* NDW, n.d.

———. *Naval Station Washington Master Plan: Washington Navy Yard/Anacostia Annex.* Washington: NDW, 1998.

Naval Historical Center. *Dictionary of American Naval Fighting Ships.* 8 vols. Washington: Naval Historical Center, 1959–1995.

———. *The United States Naval Railway Batteries in France.* Reprint. Washington: Naval Historical Center, 1988.

Naval History Division. *Civil War Naval Ordnance.* Washington: Naval History Division, 1969.

Peck, Taylor. *Round-Shot to Rockets: A History of the Washington Navy Yard and the U.S. Naval Gun Factory.* Annapolis: Naval Institute Press, 1949.

Pitch, Anthony S. *The Burning of Washington: The British Invasion of 1814.* Annapolis: Naval Institute Press, 1998.

Reilly, John C., Jr. *The Bronze Guns of Leutze Park, Washington Navy Yard.* Washington: Naval Historical Center, 1980.

———. *The Iron Guns of Willard Park, Washington Navy Yard.* Washington: Naval Historical Center, 1991.

Schneller, Robert J., Jr. *A Quest for Glory: A Biography of Rear Admiral John A. Dahlgren.* Annapolis: Naval Institute Press, 1996.

Society of Naval Architects and Marine Engineers. *Historical Transactions, 1893–1943.* New York: The Society, 1945.

Spencer, Warren F. *Raphael Semmes: The Philosophical Mariner.* Tuscaloosa: The University of Alabama Press, 1998.

Sprout, Harold, and Margaret Sprout. *The Rise of American Naval Power, 1776–1918.* Rev. ed. Princeton, NJ: Princeton University Press, 1942.

Still, William N., Jr. *Ironclad Captains: The Commanding Officers of the USS Monitor.* Washington: National Oceanic and Atmospheric Administration/GPO, 1988.

U.S. Congress. House. *Official Records of the Union and Confederate Navies in the War of the Rebellion.* 54th Cong., 2d sess. Washington: GPO, 1896. H.Doc. 40. See series 1, vol. 4, index under "Navy Department-Navy Yard Washington, Commandant of."

———. Committee on Armed Services. Subcommittee on Military Installations and Facilities. *Proposed Washington Navy Yard Construction: Hearing Before the Military Installations and Facilities Committee of the Committee on Armed Services, House of Representatives.* 97th Cong., 1st sess., October 5, 1981. Washington: GPO, 1981.

U.S. Congress. Senate. Committee on Armed Services. Subcommittee on Military Construction. *Navy's Plan to Move from Leased Space to the Washington Navy Yard: Hearing Before the Subcommittee on Military Construction of the Committee on Armed Services, United States Senate.* 97th Cong., 1st sess., September 30, 1981. Washington: GPO, 1982.

U.S. Navy. Bureau of Yard and Docks. *Annual Report.* Bound with *Annual Report of the Secretary of the Navy* (title varies). Washington: GPO, 1842–1940.

Washington Navy Yard. *Yard Log.* Washington: The Yard, 1943–1961.

Articles, Unpublished Dissertations, and Excerpts

Akers, Regina T. "Freedom without Equality: Emancipation in the United States, 1861–1963." In *First Freed: Washington, D.C., in the Emancipation Era,* edited by Elizabeth Clark-Lewis. Washington: A.P. Foundation Press, 1988.

Anthony, Susan B. II. "Working at the Navy Yard." *The New Republic,* 1 May 1944. Reprinted in *Reporting World War II,* vol. 2. New York: Library of America, 1995.

Anzelmo, Robert J. "Post with a New Name." *Washington Star Magazine,* 4 October 1959.

Bartlett, Tom. "Marines of Building 58." *Leatherneck,* August 1988.

Bureau of Medicine and Surgery, Historian's Office, "Washington, D.C." February 1999.

Daverede, A. J. "From Broadsides to BRAC: A Study in Bicentennial Survival of the Washington Navy Yard." Unpublished research paper for American Military University, 1998.

Fitzgerald, Oscar P. "History of the Washington Navy Yard." Naval Historical Center, n.d.

Glasow, Richard Dwight. "Prelude to a Naval Renaissance: Ordnance Innovation in the United States Navy During the 1870s." Ph.D. diss., University of Delaware, June 1978.

Guide to the Marine Corps Historical Center. n.d.

Guttridge, Leonard F. "Identification and Autopsy of John Wilkes Booth: Reexamining the Evidence." *Navy Medicine,* January–February 1993.

Hughes, Christine. "The Early History of the Washington Navy Yard from Its Founding in 1799 through the War of 1812." Washington: Naval Historical Center, 1997.

Jensen, Milinda. "Yard Plays Host to Historical Exchange." *Sea Services Weekly,* 4 May 1990.

Leahy, W. D. "Early History of the Washington Navy Yard." U.S. Naval Institute *Proceedings,* October 1928.

Linder, B. R. "Naval Tradition Captured: The Washington Navy Yard shows history where history was made." *Military History,* 1989.

Lobdell, George H. "USS Sequoia." *Naval History,* March–April 1998.

"Navy Band to Set Out on New Tour on Return." The *Evening Star* [Washington], 26 February 1960.

"The Navy Buries 14 Dead: 'We Give Back to the Earth...'" *The Washington Daily News,* 9 March 1960.

"Old Gun Factory Boomed After World War." *The Washington Post,* 6 April 1961.

Public Affairs Office, Naval District Washington. "The Washington Navy Yard: A Brief History." October 1968.

"Remarks at the Commissioning of the Research Ship Oceanographer, July 13, 1966." *Public Papers of the Presidents of the United States: Lyndon B. Johnson: Containing the Public Messages, Speeches, and Statements of the President, 1966.* Book II. Washington: GPO, 1967.

Roe, E. W. "Brief Historical Sketch of the Navy Yard at Washington, D.C." In *Historical Transactions, 1893–1943.* New York: Society of Naval Architects and Marine Engineers, 1945.

"A Salute to the Navy: The Washington Navy Yard." *APWA Reporter* [Public Works Historical Society], October 1975.

Semmes, Katherine Ainsworth. "A Historic Heritage: The Washington Navy Yard." Washington: Naval Officers Wives Club of Washington, D.C. N.d.

United States Navy Band. "History of the United States Navy Band." February 1999.

Index

A
ABCD ships, 39
Adams, John, 1, 8, 12, 13, 24, 89
Adams, John Quincy, 12, 24
Advanced Fire-Control School, 73
Alabama, CSS, 19, 35
Alaska, 61
Alexandria, Virginia, 23, 24, 55
America, 64
American Red Cross, 54, 55
American Revolution, 1, 13, 86
Anacostia Naval Air Station, 49
Anacostia River, 1, 18, 35, 49–50, 65–67, 72, 75, 82, 90, 97–98. *See also* Eastern Branch
Anderson, George W., 82
Andrews, E. F., 2
Annapolis, Maryland, 1, 17, 48, 88
Antarctica, 78,
Appomattox, Virginia, 32
Argentine Navy, 50
Argus, 9, 10
Arizona (BB 39), 53,
Arlington National Cemetery, 46, 79, 83
Arlington, Virginia, 89
Army Department, 33, 34
Army Surgeon General, 33
Army War College, 50
Arthur, Chester A., 39
Atlanta (CL 23) 38, 39
Atlantic Fleet, 46
Atlantic Ocean, 61
Augusta (CA 31), 65
Australia, 47
Axis Powers, 74

B
Bainbridge, William, 5
Baldwin Locomotive Works, 55
Baltimore, Maryland 9, 10, 25, 39
Baltimore, 23
Barbary Wars, 1, 3, 5, 8, 90, 93
Barney, Joshua, 9
Barry, Thomas, 25
Barry (DD 933) 85, 97
Base Closure and Realignment Commission (BRAC), 98–99
Battle of Bladensburg, 9
Battle of Brandywine, 13
Battle of Gettysburg, 32
Battle of Hampton Roads, 29
Bayless, Raymond, 28
Bellinger, Patrick N. L., 49
Benton, Thomas Hart, 24, 85
Bermuda, 9
Bethlehem Iron Company, 39
Birmingham, 47, 52
Bladensburg, Maryland, 9, 24
Blandy (DD 943), 79, 80
Board of Navy Commissioners, 11, 14
Bolles, John A., 35
Boorda, Jeremy, 95
Booth, John Wilkes, 32–34
Booth, Mordecai, 10
Boston, 39
Brady, Mathew, 19, 30–31
Brandywine, 12, 13
Brazil, 83
Brittain, Carlo B., 48
Brown, Almira V., 61
Brown, John, 21
Buchanan, Franklin, 20–21, 27
Building 1, 6–7
Building 57, 95
Building 58, 79, 95
Building 76, 71, 83. *See also* Navy Museum, The
Building 112, 82
Building 157, 72
Building 208, 32
Building 220, 69
Bull Run, 23
Bureau of Aeronautics, 59
Bureau of Construction and Repair, 57, 58, 59, 61
Bureau of Ordnance, 19, 32, 35, 37, 47, 55, 69, 82
Bureau of Ordnance and Hydrography, 17, 18
Bureau of Yards and Docks, 14, 15
Burke, Arleigh, 82, 83
Butler, J. M., 22
Buzen, Shinmi, 20
Byrd, Richard E., 78

C
Callaghan, Daniel J., 68
Cammilliari, N., 12
Camp Good Will, 61
Canada, 9
Cape Horn, 47

Capitol, 1, 13, 24–25, 27, 29, 83
Capitol Hill, 1, 25,
Carderock, Maryland, 61
Caribbean, 39,
Carter, Jimmy, 65, 94
Ceremonial Guard Company, 79
Chambers, Washington I., 48
Chandler, William E., 39
Charleston, South Carolina, 19, 21, 32
Charlottesville, Virginia, 2
Chesapeake, 4–5
Chesapeake Bay, 9, 22, 65
Chicago, 39
Chief of Naval Operations, 3, 6, 78, 82–83, 89, 94–95, 98
China, 47
Churchill, Winston, 65
Civil War, 15, 17, 19, 21–34, 91, 92
Cleveland, Grover, 42
Clinton, Hillary Rodham, 87
Cockburn, George, 9
Cold War, 75, 77–80, 85, 88, 95, 98
Colombia, 82
Colt, Samuel, 15
Columbia, 9, 10
Columbus, 10, 12
Combat Art Gallery. *See* Navy Art Gallery
Commandant Naval District Washington, 6–7, 88
Commandant's House, 3, 10, 13, 22, 32, 37.
    See also Quarters A; Tingey House
Congress, 3, 8, 27
Congress, U.S., 1–4, 11–12, 27, 36–37, 45, 60, 66–67
Connally, John B., 83
Constellation, 3, 8
Constitution, 1, 3, 8, 13, 67, 82–83, 86, 89–90, 94
Constitution Avenue, 89, 94
Continental Navy, 1
Coolidge, Calvin, 61–62, 64
Country Current, 95
Cozzens, Fred S., 5
Crowninshield, Benjamin, 11
Cruzen, Richard H., 78
Cumberland, 27
Cunningham, A. A., 48
Currier, N., 15
Curtiss, Glenn H., 48
Curtiss A-1, 48
Curtiss Pusher plane, 47, 48, 52
Curtiss seaplane, 52

D

Dahlgren, John A., 7, 15–23, 27–29, 32, 37, 72, 90–91
Dahlgren, Mary Clement, 28
Dahlgren, Ulric, 28, 29
Dahlgren Avenue, 91
Dahlgren gun, 16, 18–19

Daniels, Josephus, 55, 56, 92
Danzig, Richard, 87
David Taylor Model Basin, 60, 61.
    See also Experimental Model Basin
Decatur, Stephen, 5
Deep Diving School, 59
Defense Department, 94, 98
Defense Department Computer Institute, 94
Despatch, 39, 46, 64
Dewey, George, 45
Director of Naval History, 87, 93, 95
District of Columbia, 2, 6–7, 24–25, 35, 42, 86–88, 94, 99
District of Columbia Emancipation Act, 27
Dolphin, 39, 46, 64
Dreadnought, 46
Dudley, William S., 91
Dudley Knox Center for Naval History, 93, 95.
    See also Naval Historical Center
Durand, Asher B., 12

E

Eagle, 84
Earle, Ralph, 55
Early, Jubal, 29
East Coast, 1, 65, 79
Eastern Branch, 2, 11, 15, 17, 20, 24.
    See also Anacostia River
Egypt, 35
Eighth Street, 7
Eisenhower, Dwight D., 65, 80, 94
Electric Hydraulic School, 73
Ellsworth, Elmer Ephraim, 23
Ellyson, Theodore G., 47, 48
Ely, Eugene, 47, 48
Emperor Hirohito, 94
England, 35, 46, 64, 91
Enterprise, 8
Environmental Protection Agency, 99
Ericsson, John, 29
Europe, 53, 66–67, 73, 93
Evans, Robley D., 47
Executive Office Building, 92
Experimental battery, 16–20, 25, 72, 90
Experimental diving unit, 59
Experimental Model Basin, 41–43, 45, 48–49, 54 56–59

F

Far East, 66
First Battle of Bull Run, 23
1st Zouave Regiment, 23
Flood, 59, 66
Florida, 21, 49, 65
Flying Squadron, 45
Ford, Gerald, 94

Forrestal, James, 78, 93
Forrest Sherman (DD 931) class, 85
Fort Ellsworth, 23
Fort Leslie J. McNair, 9, 34
Fort Lincoln, 29
Fort McHenry, 10
Fort Monroe, 27
Fort Sumter, 21

G
General Electric, 45
General Services Administration, 82
George K. MacKenzie (DD 836), 73–74
Georgetown, 1–2, 13, 24
Germany, 49, 66, 72
Graham, John R., 87
Grant, Ulysses S., 36, 64
Great Britain, 1, 5, 9, 20, 39, 59, 65, 67–68
Great Depression, 64, 66
Great War, 57
Great White Fleet, 46, 47, 51
Greece, 82
Greek Revival, 6
Greenleaf Point, 9–10
Gross, J., 9
Guardian, 85
Guerriere, HMS, 13
Gulliver, Louis J., 67
Gun Foundry Board, 39
Gunnery School, 73

H
Hamilton, Paul, 5
Hampton Roads, 27–29, 46–47, 51
Harding, Warren G., 49, 61
Harpers Ferry, West Virginia, 21, 29
Harrison, John, 35
Harwood, Andrew A., 32
Havana Harbor, 44
Henry Hill, 23
Hill, Andrew J., 88
Historic District, 94
Honey Fitz, 85
Hoover, Herbert, 64, 67, 94
Hornet, 8
House of Representatives, 12
Hull, Isaac, 13, 24, 27, 41–43, 57, 59, 61
Hunt, William H., 39
Huntress, 3

I
Ignatius, Paul R., 88
Illinois, 34
Imperial Japanese Navy, 54

Independence Day, 45
Indian Head, Maryland, 56
Iowa (BB 61) class, 58, 69
Israel, 88–89
Italy, 35, 39, 59, 66

J
Jackson, Thomas "Stonewall," 29
Japan, 20, 47, 59, 66, 72, 86, 90
Jaques, William, 39
Jeffers, William N., 35, 37
Jefferson, Thomas, 2–3, 5–6
John Fritz Medal, 57
Johnson, Lyndon B., 88
Joint Chiefs of Staff, 89
Joint Committee on Landmarks, 94
Judge Advocate General, 35, 98

K
Kalbfus, Edward C., 93
Kamikaze, 71
Kansas (BB 21), 61
Kane, John D. H., Jr., 95
Kearsarge, 19
Kennedy, John F., 65, 84–85, 87
Kettlebottom Shoals, 28
King George VI, 65, 67, 68
King, Thomas, 66
Knox, Dudley W., 92–93
Korean War, 79, 83
Korth, Fred H., 87
Kuwait, 81

L
Lafayette, Marquis de, 12–13
Land, Emory S., 59, 61
Langley (CV 1), 49, 63
Langston, Arthur N., 87
Laon, France, 56
Latrobe, Benjamin Henry, 2–3, 6–7
Latrobe Gate, 3, 6–7, 20, 21, 22, 63, 65, 95, 97
Leopard, HMS, 4, 5
Leutze, E.H.C., 45
Leutze Park, 56, 74, 79, 82, 90, 94–95, 96–97
Liberty (AGTR 5), 88–89
Liberty Bonds, 55
Lincoln, Abraham, 19, 21, 23, 26, 27–29, 32–35
Lindberg, Charles A., 61, 63
London, 63, 66, 92
London Naval Treaty, 63, 66
Long, John D., 45
Lord Kelvin, 57
Lovering and Dyer, 2, 6–7
Lynx, 10

M

Macedonian, HMS, 11
Madison administration, 9
Mahan, Alfred Thayer, 45
Main Navy, 89, 94
Maine, 23, 44, 46
Manila Bay, 45
Marine Corps Historical Center, 94, 95, 98
Marine Parade, 75
Marsh, Reginald, 83
Massachusetts Avenue, 94
Mayflower (PY 1), 46, 50, 61, 64–65, 80
McDonald, David L., 82
McGonagle, William L., 88
McKinley, William, 64
McNamara, Robert S., 87
Medal of Honor, 68, 88
Medical Department, 73
Mediterranean, 88
Meiji Restoration, 20
Memorial Bridge, 81, 83
Memphis (CL 13), 61, 63
Merrimack, 17, 27
Metz, France, 56
Mexican War, 12, 14
Mexico, 82
Miantonomoh, 33
Military Sealift Command, 94
Mine Disposal School, 73
Minnesota, 14, 27
Missouri (BB 63), 74, 75
Mohammad Ayub Khan, 85
Monitor, 19, 28–29, 33, 35
Monroe, James, 12
Montauk, 33
Monticello, 2
Morison, Samuel Eliot, 93
Morris, Charles, 17
"Mosquito Squadron," 5
Mount Vernon, 22–23
Mount Vernon estate, 65, 67, 85
Murray, Erick Marshall, 88

N

Napoleon, 5, 90
"Nasty" class, 45, 87
National Register of Historic Places, 7, 94
Naval Act of 1916, 53–54
Naval Battery, 23
Naval Dental Clinic, 94
Naval District Washington, 6–7, 88, 99
Naval Facilities Engineering Command, 94, 98
Naval Gun Factory, 18, 39–40, 45–46, 48-49, 53, 56, 59, 66, 69, 74–75, 82–83, 86
Naval Historical Center, 7, 94–95, 98
Naval History Division, 89, 94
Naval Investigative Service, 94
Naval Observatory, 94
Naval Ordnance Laboratory, 73
Naval railway battery, 55–56, 58
Naval Records and Library, 92–94
Naval Regional Data Automation Command, 94
Naval Reserve, 54, 93
Naval School of Diving and Salvage, 77
Naval Sea Systems Command, 99
Naval Station Washington Master Plan, 99
Naval War Records, 92
Navy Art Gallery, 85
Navy Cross, 92
Navy Department, 2–3, 13, 17, 23, 27, 32, 34–36, 45, 48, 55, 61, 93
Navy Department Library, 89
Navy Museum, The, 82–83, 86–87, 91, 98
Navy Yard Apprentice School, 73
Navy Yard Hill, 4–5
Nevada, 53
New England, 46, 64
New Jersey (BB 62), 58
New York, 3
New York, 46, 64
New Zealand, 47
Newport News, Virginia, 27
Nimitz, Chester W., 78
Nippon Maru, 82
Nipsic, 37
Nixon, Richard M., 65, 94
Norfolk, Virginia, 5, 17, 22, 87
Norfolk Navy Yard, 22
North Africa, 1, 5, 93
North Atlantic, 53
North Atlantic Squadron, 45
North Carolina (BB 55) class, 69
North Sea Mine Barrage, 54, 56
North Vietnam, 87, 94
Northern Virginia, 89, 98
Norton, Eleanor Holmes (D–D.C.), 87
Norway, 54, 87

O

Office of Naval Records and Library, 92–93
Ohio Railroad, 39
Oklahoma, 53
Olympia, 58
Operation Desert Storm, 81
Optical Primary School, 73
Ordnance Department, 22
Ordnance Establishment, 18, 20
Ordnance Instruction for the U.S. Navy, 37
Ordnance School, 73

P

Pacific, 19, 39, 71, 85–86
Paraense, Brazilian, 31
Paris, 55, 56
Patrick J, 85
Patuxent River, 9
"Peacemaker" gun, 15, 91
Penn Bridge Company, 42
Pennsylvania, 53
Pennsylvania Avenue, 20, 84
Pensacola, Florida, 21, 49
Perry, Matthew C., 20, 86, 90
Philadelphia, 20, 65
Philippines, 47
Poland, 69
Polk, James K., 17
Port Authority, 95
Porter, David Dixon, 20, 36
Portsmouth, 46
Potomac, 12
Potomac (AG 25), 63, 64–65
Potomac Flotilla, 22–23, 35, 36
Potomac River, 1, 9, 11, 22, 24, 46, 64, 87–88, 91
Potomac River Naval Command, 88
Power Boat Regatta, 75
President, 3
Princeton, 15, 25, 91
privateers, 1, 3, 5
PT 109, 84, 85
PT 796, 84, 85
PTF 3, 87

Q

Quarterdeck, 79, 99
Quarters A, 3, 6–7. *See also* Commandant's House; Tingey House
Quarters B, 6–7
Quasi-War with France, 2, 93
Queen Elizabeth, 65, 67, 68

R

Raby, J. J., 62
Radical Republicans, 36
Reconstruction, 35
Resolute, 23
Richardson, Holden C., 48–49
Richmond, 32
Robison, Samuel S., 48
Rodgers, John, 12
Roosevelt, Eleanor, 68
Roosevelt, Franklin D., 48, 61, 63–69, 93
Roosevelt, Theodore, 9, 45–46, 51, 64
Ross, Robert, 9
Royal Navy, 2, 5
Russia, 20, 39
Russo-Japanese War, 46, 64

S

Sail Loft, 45, 82
Sampson, William T., 45
San Francisco, California, 47, 61
Sandy Hook, New Jersey, 56
Santiago de Cuba, 45
Sarmiento, Argentine, 50
Saugus, 33
Schley, Winfield Scott, 45
Schmidt, John W., 8
Scimonelli, Frank J., 83
Scotland, 54, 64
Sea Chanters, 95
Secretary of State, 15
Secretary of War, 27–28, 33–34, 39
Semmes, Raphael, 34–35
Sequoia, 63-65, 85, 89
71st New York Infantry Regiment, 7, 22–23
Severn River Naval Command, 88
Shiner, Michael, 17, 24–25, 27
Simons, Manley H., 48
Sims, William S., 46, 92
Skerrett, R. G., 14
Smith, Robert T., 7
Soley, James Russell, 92
South Atlantic, 47
South Atlantic Blockading Squadron, 19
Southeast Asia, 87
Soviet Union, 79, 95, 98
Spain, 5, 45, 64, 90
"Spirit of St. Louis," 63
Springfield, Illinois, 34
Stalin, Joseph, 98
Standard Steel Car Company of America, 55
Stanton, Edwin M., 27–28, 33–34
Stockton, Robert, 91
Stoddert, Benjamin, 1–2, 89
Submarine, 53–55, 57, 59, 77, 86
"Summer Ceremony," 95
Sumter, 35
Surgeon General, 33
Surratt, John, 34–36
Surratt, Mary, 34
Susquehanna, 13
Swanson, Claude, 61
Swatara, 35
Sweden, 20
Swift boat, 85
Sylph (PY 12), 46, 64

T

Taft, William Howard, 50
Talos missile, 85
Tartar missile, 82
Taylor, David Watson, 42, 45, 56–58, 61

Taylor Standard Series, 56–57
Thrift Stamps, 54
Tingey, Thomas, xiv, 2–3, 5–10, 13, 54, 94–95, 97
Tingey House, 3, 6, 54, 94, 95, 97.
    See also Commandant's House; Quarters A
Tokyo Bay, 74
Tom O'Malley (YRST 5), 77
Tonkin Gulf Incident, 87
Towers, John H., 48
Trieste, 83, 86
Tripoli Monument, 1,
Truman, Harry S., 65, 73, 75, 79
Tyler, John, 15

U
U.S. Coast Geodetic Survey Ship *Oceanographer*, 88
U.S. Coast Guard, 84
U.S. Marine Corps, 2, 6, 79, 94–95, 98
U.S. Naval Academy, 1, 91–92
U.S. Naval Historical Display Center. *See* Navy Museum, The
U.S. Naval Institute, 36
U.S. Naval Weapons Plant, 82
U.S. Navy, 1, 18, 25, 57, 61–62, 74, 82–83, 86, 89, 95
U.S. Navy Ceremonial Guard, 95
U.S. Navy Memorial Museum. *See* Navy Museum, The
U.S. Senate, 75
Union Navy, 19, 22
United States, 3, 5
United States Naval Weapons Plant, 82
United States Navy Band, 61–62, 74, 82–83, 86, 95
Urban League, WALL program, 87
USSR, 98

V
Verdun, 56
Victory in Europe (VE) Day, 73
Vietnam War, 83, 85–87, 94
Vinson-Trammell Act, 66
Virginia, CSS, 19, 27, 28
Vixen, 8
Volkogonov, Dmitri A., 98
VT fuse, 69

W
Wabash, 28
War of 1812, 9–10, 11–12
Warrington, Lewis, 15
War Savings Certificates, 54
Washington, George, 1
Washington Naval Treaty of 1922, 59, 63
Washingtonians, 10, 32, 81
Wasp, 4, 8
Water Pageant, 75
Water Witch, 14
"Watergate Concert," 81
Watervliet Arsenal, 45
Weapons Engineering Support Activity, 94
Welles, Gideon, 21–22, 28–29, 32, 35–36
West Coast, 47
White House, 13, 23, 27–28, 65, 92
White Oak, Maryland, 73
Whitney, William C., 39
Willard Park, 83, 85, 90–91
Willard's Hotel,, 20
William Sellers Company of Washington, 45
Williams, Anthony, 87
Williamsburg (AGC 369), 65, 79
Wilson, Woodrow, 53, 64
Winder, William H., 9–10
Wisconsin (BB 64), 81
Women Accepted for Volunteer Emergency Service (WAVES), 73
Women workers 54, 61, 69–70, 72–74, 88
World War I, 30, 47, 53–56, 58–59, 61, 64, 66, 79, 90, 92
World War II, 64–65, 68–70, 72, 74–75, 78–79, 83, 85–86, 93, 95
Wright, Orville, 48, 57
Wright, Wilbur, 48
Wyoming, 90

Y
Yantic, 37
Yates, Theodore, 91

ISBN 0-16-050104-0